EMERGENCY MANAGEMENT
AND SUSTAINABILITY

ABOUT THE AUTHOR

Robert O. Schneider is a Professor of Public Administration at the University of North Carolina at Pembroke.

EMERGENCY MANAGEMENT AND SUSTAINABILITY

Defining a Profession

By

ROBERT O. SCHNEIDER, PH.D.

Department of Public Administration
University of North Carolina at Pembroke

CHARLES C THOMAS • PUBLISHER, LTD.
Springfield • Illinois • U.S.A.

Published and Distributed Throughout the World by

CHARLES C THOMAS • PUBLISHER, LTD.
2600 South First Street
Springfield, Illinois 62704

© 2013 by CHARLES C THOMAS • PUBLISHER, LTD.

ISBN 978-0-398-08763-0 (paper)
ISBN 978-0-398-08764-7 (ebook)

Library of Congress Catalog Card Number: 2013022553

With THOMAS BOOKS *careful attention is given to all details of manufacturing
and design. It is the Publisher's desire to present books that are satisfactory as to their
physical qualities and artistic possibilities and appropriate for their particular use.*
THOMAS BOOKS *will be true to those laws of quality that assure a good name
and good will.*

Printed in the United States of America
SM-R-3

Library of Congress Cataloging-in-Publication Data

Schneider, Robert O.
    Emergency management and sustainability : defining a profession / by
Robert O. Schneider, Ph.D., Department of Public Administration, University of North Carolina at Pembroke.
        pages cm
    Includes bibliographical references and index.
    ISBN 978-0-398-08763-0 (pbk.) -- ISBN 978-0-398-08764-7 (ebook)
    1. Emergency management--United States. I. Title.

HV551.3.S376 2013
363.34'802373--dc23

                                                                    2013022553

PROLOGUE: SOMETHING ELSE IS NEEDED

The footprints of disaster are bigger and costlier than ever. The past few years, if you think about them for any length of time, do make that impression. The first decade of the twenty-first century has presented a host of new or different challenges and made normally recurring challenges more complex. A seemingly unending series of more dramatic large-scale natural disasters, the worldwide terrorism threat, the possibility of international influenza pandemics, the threat of a worldwide cyber-failure, the potential impact of global climate change, the Haitian earthquake of 2010, the Fukushima nuclear disaster of 2011, Superstorm Sandy in the fall of 2012 dramatically impacting the densely populated American Northeast, and the environmental threats that are frequently posed by promising new technologies that frequently create new risks for humanity and threaten our natural resources (e.g., natural gas fracking, deep water oil drilling, etc.) are just a few of the recent experiences and concerns that may cause us to wonder whether we are living in an era in which disasters (natural and human-made) and the damages they cause might be beginning to exceed our capacities to manage effectively.

Six of the top ten natural disasters in the past one hundred years, as measured in terms of lives lost and property damage, have taken place since 2001 (i.e., listed in order of severity: Haiti Earthquake, 2010; Indian Ocean Earthquake/Tsunami, 2004; Cyclone Nargis, 2008; Japan Earthquake/Tsunami, 2011; Gujarat Earthquake, 2001; and Hurricane Katrina, 2005). Human-made hazards resulting in disasters have also made some more dramatic and historically significant appearances in the first decade of the twenty-first century (e.g., Al-Mishraq Sulfur Fire in Iraq, 2003; Jilin Chemical Plant Explosions in China, 2005; and the BP Oil Disaster in the Gulf of Mexico, 2010). Each dramatic event seems to invite renewed assessment of our ability to be resilient in the face of the array of natural and human-made hazard threats that hold the potential to bring bigger and more destructive disasters to our doorsteps. The rising costs associated with each event also invite our attention.

The number and costs of major natural disasters in the United States, for example, are on the rise. From 1980 through 2012, 144 natural disasters in the

United States caused damages in excess of $1 billion (Table 1). But 25 out of these 144 have occurred in the last two years, 2011 and 2012.

Table 1. 1980–2012 U.S. Natural Disasters Exceeding $1 Billion in Damages

Year	$Billion+ Disasters	Year	$Billion+ Disasters
1980	2	1997	3
1981	1	1998	9
1982	1	1999	5
1983	4	2000	2
1984	1	2001	2
1985	5	2002	3
1986	1	2003	5
1987	0	2004	5
1988	1	2005	5
1989	4	2006	6
1990	3	2007	5
1991	3	2008	9
1992	6	2009	6
1993	4	2010	4
1994	6	2011	14
1995	4	2012	11
1996	4	Total	144

Source. National Oceanic and Atmospheric Administration.
http://ncdc.noaa.gov/billions/events/pdf

According to the National Oceanic and Atmospheric Administration, eleven extreme weather or climate-related events in 2012 caused damages exceeding $1 billion in the United States (Table 2). This is three fewer than in 2011, but the aggregate costs of the 2012 billion dollar events are expected to exceed those of 2011. The eleven billion dollar plus events of 2012 include seven triggered by severe weather or tornadoes, two hurricanes, and two others resulting from the impact of the extreme drought that gripped much of the nation throughout the year.

Estimates at the beginning of 2013 are that 2012 will be the second most costly year for natural disasters (2005 is number one) in the United States in the 1980–2012 timeframe. The two events that were the biggest drivers of costs in 2012 were Superstorm Sandy (60+ billion) and the yearlong drought (40+ billion). The 2012 drought conditions were the worst in the United States since the 1930s. They impacted more than half the country for a majority of the year. Many experts are predicting that billion dollar weather or climate events will become the norm in the years to come.

Table 2. Billion+ Dollar Natural Disasters in 2012 in the United States

Date	Event	Cost Estimates ($)*
March 2–3, 2012	Southeast/Ohio Valley tornadoes	4+ billion
April 2–3, 2012	Texas tornadoes	1.3 billion
April 13–14, 2012	Midwest tornadoes	1.75 billion
April 28–May 1, 2012	Midwest/Ohio Valley severe weather	3+ billion
May 25–30, 2012	Southern Plains/Midwest/Northeast severe weather	2.5 billion
June 6–12, 2012	Rockies/Southwest severe weather	1.6 billion
June 29–July 2, 2012	Plain/East/Northeast severe weather	3.75 billion
August 26–31, 2012	Hurricane Isaac	3 billion
Summer–Fall 2012	Western wildfires	na
October 29–31, 2012	Superstorm Sandy	60–65 billion
Throughout 2012	U.S. drought/heat wave	40+ billion

Source. National Oceanic and Atmospheric Administration.
http://ncdc.noaa.gov/billions/events/pdf
*damage estimates are derived from a variety of public sources, and some estimates are preliminary.

The United States and countries around the world have, of course, extensive experience with natural and industrial disasters. They have made significant and expanding efforts over time and committed significant resources to disaster preparedness, disaster response, disaster recovery, and disaster mitigation. But for all of these efforts and all that has been learned and done, for all of the progress made, recent events have begun to suggest that adequate risk reduction measures (i.e., mitigation) and disaster preparedness (i.e., capacity to respond) are, if not in fact declining, at least lagging behind. Economic and insured losses from natural disasters have increased steadily over the years, and they show every indication of continuing to escalate. This is due not only to the occurrence of damaging natural events, which are variable from year to year but are by many indications intensifying in severity, but also to some basic demographic factors. Over half of the U.S. population, for example, now resides in coastal counties (30% on coastlines bordering the ocean or associated water bodies). This number has risen steadily and is expected to continue rising. This shift places more people and more expensive development in high-risk areas and, inevitably, increases the impact and economic and insurance losses associated with tropical events. It must also be noted that the anthropogenic or human-made hazards that are inevitably the result of technological and industrial development are always an ever-present threat to be managed and that these too seem to be expanding with our continued progress.

Changing natural hazard patterns, development strategies and policies, changing demographics, and changing economic conditions contribute to changing risk and vulnerability profiles in relation to hazard threats and potential disaster impacts. It can be difficult to keep up with hazard risks and disaster-related concerns. Emergency managers are faced with natural and human-made problems that are constantly evolving and changing the footprints of disaster. The complexity of these problems is more than matched by the complexity of the physical and social systems that emergency managers are expected to understand as they offer solutions for the recurring disaster problems that are presented to them in the normal course of their work. The technical skills and capacities that emergency managers have developed over time as they have plied their trade are impressive and increasingly effective. But they are not nearly enough to keep pace with or manage hazard risks and disasters. Something else is needed.

During the 1990s, the themes of hazard mitigation, hazard resilience, and sustainability became prominent in the emergency management literature. The need to assess and manage risks and vulnerabilities and to take steps to promote hazard resilience, the connectivity of risk assessment and risk management to environmental sustainability, and the urgent need for new thinking about human communities and the ever-changing threats to their environmental, economic, structural, and social sustainability were all emphasized and to some extent represented the "something else" that was needed to successfully manage risks and reduce the negative impacts of disasters. But a synthesis never really developed to establish this growing awareness as a centerpiece for defining the broader role of emergency management and the work of its practitioners.

The discussion of hazard resilience and its relationship to sustainability actually invites the integration or mainstreaming of emergency management into the sustainability framework as a necessary component. But to capitalize on that invitation, it is first necessary to have a conceptualization of emergency management that goes beyond its technical skills and specific functions. There is a need, in other words, for a worldview that is built on the connections among hazard threats, disaster or hazard resilience, and sustainability. This worldview must begin with the broad realization that a sustainable development framework requires a clear prescription and a practical application of disaster management and the effort to reduce disaster risks.

The purpose of this book is to define emergency management as a profession, something that has been discussed much in recent years but not brought to a satisfactory completion. The linkage of emergency management to sustainability, the defining of it as a sustainability profession, is presented herein as the necessary ingredient that holds the potential to orient all of the professional skill development and the work of the "trade" and to transform it into a "profession." This transformation, the "something else" if you will, is a necessity to assure ourselves that disasters (natural and human-made) will never exceed our capacities to manage effectively. This transformation, which if successfully completed better enables whole communities to take responsibility for disasters, is needed to promote hazard resilience in particular and sustainable communities in general.

An examination of the functions and strategies that occupy emergency management practitioners on a daily basis suggests broadly to academics and practitioners alike that they are aligned instrumentally to the concepts of sustainability and sustainable development. But that alignment has never served as the foundation for what emergency managers do, as the definition of their role if you will. The connection of emergency management to sustainability and the importance of sustainable risk management are not new themes. They are themes that have stimulated ongoing analysis and discussion in the emergency management literature. They are themes that grow more important by the day in fact, so much so that the time has come to define emergency management as a sustainability profession.

ACKNOWLEDGMENTS

Several important people have played a role in assisting (and lessening the burdens for) this author during the past several months. First, I must acknowledge my graduate assistant, Samantha ("Sam") Cornwell. Her assistance in performing the varied tasks assigned to her in connection to this effort has made me remarkably more efficient and productive. Second, a special thank you is owed to my administrative assistant, Amelia K. Elk, who likewise expedited progress on this manuscript by good naturedly and efficiently incorporating into her duties a number of tasks regarded as "urgent" by a harried author and that paved the way for me to be efficient in this project. Finally, and most important, I would like to acknowledge my wife, Doris. Her confidence and support, her considerable skill as a proofreader, her honest input, her enthusiasm for the subject matter, and her saintly patience with a preoccupied and somewhat distracted husband were essential contributions without which this manuscript could not have been completed.

CONTENTS

EMERGENCY MANAGEMENT
AND SUSTAINABILITY

Chapter 1

EMERGENCY MANAGEMENT: TRADE OR PROFESSION

INTRODUCTION

Every emergency management practitioner and scholar has heard or told a variation of the same joke. Asked by a county commissioner to describe what emergency managers do, the county Director of Emergency Management says, "My job is to tell you things you don't want to hear, asking you to spend money you don't have, for something you don't believe will ever happen" (Whitaker, 2007). This humorous description generally produces a friendly chuckle or two, but it should also produce more than just a little concern within the emergency management community.

For all the work emergency managers do (preparing their communities for disasters, organizing the response to them, mitigating disaster impacts, or assisting in the recovery from them), few in the communities they serve know much about who they are and what they do, including, unfortunately, many elected officials. But when a disaster is imminent, when the spam is ready to hit the fan so to speak, everyone is generally pleased that emergency managers (whatever the heck they are) are on the job. This is due in part to the fact that most people do not think about disasters, natural, industrial, or any other, until they are happening or just about to happen. Unless it is imminent, a disaster is of low salience to most people most of the time. When their attention is elevated, the immediate impact and response phase grabs their attention, and that is often assumed to be the primary job of emergency managers. This misperception, of course, is common. First response, while a technical and important function, should not be confused with what is meant by emergency management. Common misperceptions aside, there is perhaps a more significant reason that people do not quite recognize what emergency management is or what its practitioners do. This has to do with what might be called, for lack of better terminology, its lack of professional standing.

A "profession" is, generally, understood by a broader public because it has recognizable characteristics that everyone more or less comprehends. A profession typically has credentialing and certification requirements. These include things such as advanced education requirements and licensure. A profession controls its "professional standing" by setting and enforcing standards that guide individual performance and govern professional advancement. Formal accreditation is frequently a method by which such standards are enforced. But emergency management has not yet evolved to where these characteristics have been fully developed. While some efforts are being made to advance emergency management toward a profession (i.e., through initial discussions and important first steps promoting educational requirements or training, and progress toward accreditation and certification), it is not there yet.

Discussions in the emergency management literature make the case for emergency management as a profession (Crews, 2001), discuss its potential to become a profession (Lindell, Prater, and Perry, 2007), and analyze the ongoing efforts at professionalizing it (Oyola-Yemaiel and Wilson, 2005). There are many more discussions about what it should become or what its future should be with respect to education and training or its autonomy as a profession (Clement, 2011; Cwiak, 2011; Haddow and Bullock, 2005; Moore, 2010). These discussions, which have been ongoing for the past two decades, suggest that emergency management is evolving toward something but has not yet fully traversed the distance required to close the gap between trade and profession. But to truly understand where it is and where it needs to go, we might benefit from first briefly looking at where emergency management has been.

THE EARLY DAYS OF EMERGENCY MANAGEMENT

Emergency management as it is discussed and understood today consists of the work related to four disaster phases: disaster preparedness, disaster response, disaster recovery, and disaster or hazard mitigation.

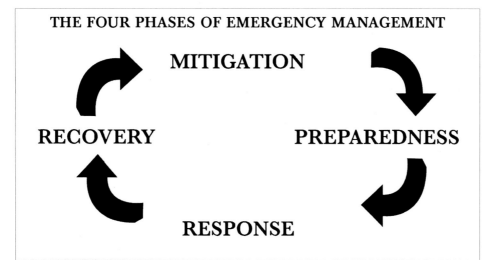

THE FOUR PHASES OF EMERGENCY MANAGEMENT

MITIGATION

RECOVERY   PREPAREDNESS

RESPONSE

Preparedness: Any set of activities that prepares for, such as creating plans, capacities, and technologies/equipment to respond to disasters (natural, technological, etc.) that a community may reasonably expect to experience on a recurring basis.
Response: Actions (i.e., the implementation of preparedness plans) taken in response to a disaster occurrence to save lives, assist victims, prevent further damage, and reduce the effects of the disaster.
Recovery: Actions taken to return to a normal or an even safer situation following the crisis.
Mitigation: Actions taken to reduce the impact or decrease the likelihood of serious losses in a future disaster scenario or, in some cases, to reduce the likelihood of disaster occurrence.

Most public attention is focused on disaster response and recovery. This is due to the fact that, once again, most people give little thought to natural or human-made disasters until they happen. But emergency management has evolved into something far beyond disaster response, although that may be said to be where it began.

An examination of the emergency management literature suggests that, until the mid-1990s, the strategic motivation of what might theretofore have been implied by the label emergency management in the United States arose mostly from the challenges of responding or reacting to specific and immediate disasters. Local communities have of course always had to deal with the impact of recurring natural disasters (floods, wild fires, tropical storms, etc.), and over time and repeated experiences, the development of both capacities and equipment to respond to these events has evolved accordingly, albeit inconsistently, across the country. The federal role in emergency management evolved beginning in the 1930s when the Reconstruction Finance Corporation was authorized to make disaster loans for the repair and reconstruction of public facilities after disaster occurrences, the Bureau of Public Roads was authorized to fund the repair of highways damaged by national disasters, and the U.S. Army Corps of Engineers was given greater authority to implement flood control projects. During the 1950s and the early 1960s, the height of the Cold War, national security and civil defense concerns dominated the emergency management agenda of the federal government and of the general public as well. This time period included the creation and implementation of federal grant initiatives and incentives to promote civil defense preparedness as a priority at the state and local levels. By the late 1960s and the 1970s, however, massive and more frequently occurring natural disasters began to result in major increases in federal disaster relief and recovery operations to assist states and localities (Federal Emergency Management Agency [FEMA]). This would lead to a necessary consolidation and restructuring of national efforts.

As federal disaster assistance expanded, and as hazards associated with the transportation of hazardous materials and nuclear power plants were added to the federal role in natural disaster planning, relief, and assistance, more than 100 federal agencies were soon involved in some capacity in the federal role in relation to natural and industrial disasters. At the urging of the nation's governors, who reasonably objected to the maze of different agencies they had to interact with to access federal assistance, President Jimmy Carter consolidated federal emergency functions with an executive order creating FEMA. FEMA absorbed the various federal disaster assistance and relief functions, along with civil defense responsibilities, and developed an Integrated

Emergency Management System. It adopted an all hazards approach. This meant the coordination of the federal role in the full range of natural and industrial emergencies from isolated incidents to major regional disasters to national incidents, including war (FEMA).

Through the 1980s, disaster response was largely thought of as a local function. The federal role was to provide assistance primarily in the areas of predisaster preparedness and disaster recovery. But even in regions or areas with a long history of regularly occurring natural disasters, emergency management issues were of low salience in most states and local communities (Wolensky and Wolensky, 1990; Wright and Rossi, 1981). In fact, the literature often noted indifference or outright opposition to disaster preparedness by many local governments (Kreps, 1991). Public officials and the communities they served were not, it seems, interested in disasters of any sort unless they were actually happening or about to happen. It is also clear that they did not fully comprehend the emergency management function beyond the need for the development of first response capacities. A basic assumption they made, and it is still made by some to this day unfortunately, was that emergency management is primarily a "response" function and only a concern for first responders. The need for other officials, elected or unelected, and the public to be involved in predisaster planning or preparedness was assumed not to exist (Grant, 1996). The functions associated with disaster recovery and hazard mitigation were likewise nothing more than little noticed blips in most communities across the country. After all, the federal government would always fund disaster relief, and mitigation was little more than a new notion at the time. Local efforts such as they were focused primarily on disaster preparedness and first response.

The full development of the emergency management function at the local level was bolstered only by federal legislation such as the Emergency Planning and Community Right to Know Act of 1986 (EPCRA) and the Stafford Act of 1988. The former (EPCRA) required local businesses to report the emissions and/or releases of toxic chemicals and also required that such information be made available to the public. It also mandated local planning for the collection of such information and the preparation of community plans for responding to disasters related to chemical emissions, spills, releases, and so on. It was in conjunction with this law (EPCRA) that local community emergency management coordinators were first designated and local planning was mandated.

The Stafford Act expanded federal discretionary authority (and organized the variety of federal programs) aimed at providing, in addition to recovery assistance or disaster relief, state and local assistance for disaster preparedness and hazard mitigation. This included the 2000 extension of the Stafford Act, for the first time establishing some criteria for hazard mitigation as a condition for receiving federal disaster relief. Such legislation began establishing federal mandates requiring local efforts in disaster planning and mitigation. Yet such requirements were slow to change priorities at the local level. Unless a specific hazard was connected to a disaster that was more or less imminent, sustained governmental interest and public support were slow to materialize (Perry and Mushkatel, 1984). Policymakers and stakeholders alike tended to underestimate disaster potentials. They were inclined to see disaster occurrence as having a low probability and were thus reluctant to impose limitations on private property and development, unwilling to bear the costs of disaster preparedness, and were mostly ambivalent about hazard mitigation (Grant, 1996). Emergency management remained a low priority, a resented unfunded federal mandate to some extent, and a responsibility often seen as being somewhat at odds with higher priority tasks such as economic development. These attitudes shaped and restrained the early development of the emergency management function in local communities across the country.

Through the 1980s and into the early 1990s, the emergency management function suffered as a result of low political support and scarce resources. In many local jurisdictions, it became an added or part-time responsibility for an already overburdened local official such as a fire chief. Often the individuals appointed to local emergency directorships, largely in response to new federal requirements, had little formal training or experience relevant for the job. As a result of low salience, poor training, lack of support, and often a first responder orientation, the emergency manager tended to be narrow of perspective, disaster specific, technical, and limited to specific tasks connected to the next disaster and the preparedness for and response to it. Emergency management was, one might say, a trade. It was an occupation that required some skills, even technical skills, but it was not a distinct profession. This would begin to change somewhat in the 1990s. It is probably accurate to say that since the 1990s, the emergency management function has *begun* to move toward becoming a profession. But there has been, and there remains, a strategic tension that impedes the professionalization of the trade. What was lacking initially, what was found briefly and then lost (see Chapter 2), what optimists still see the potential for is a dynamic model for transforming a once limited function into a contemporary professional management role connected to the whole of community life.

Any effort to expedite the professionalism of emergency management will be frustrated if it is not connected to a strategic orientation that broadens the scope and impact of the trade and leads to the authoritative standing of its practitioners as professionals having the proper credentialing, certification, autonomy, knowledge, and authority to lead and act in a broader strategic arena. But the long-term institutional building capacity of emergency management has been slow to develop and uneven in its implementation. This may owe as much to the habits of its practitioners as to any other obstacles to the development of any new profession.

Many practicing emergency managers are more or less happy with a narrow, technical, and limited role confined to the performing of different technical occupational tasks. In fact, it runs contrary to the experience of most emergency managers to think outside of the narrow and technical aspects of their work. They are accustomed to and comfortable with a narrowly defined conception of their role as planning for a particular set of reactive, anticipatory, and planned responses to specific hazards, threats, and emergencies. This is important work to be sure, but a transformation and enlargement of the role is necessary if emergency management is to become a profession and to promote success in the broader and more strategic environment that has come to shape the contemporary work of emergency management. Beginning with the 1990s, in conjunction with the progress made in refining and improving the technical skills associated with the trade, emergency management was indeed beginning to be relocated in a wider and more dynamic context. This imposed new challenges and presented new opportunities. Some of the first signs of possible movement toward professionalizing became more evident as a result.

The literature during the 1990s and into the new century began to talk about a different emergency management, one not confined to preparing for, responding to, and recovering from specific disasters. Increasingly, emergency management came to be seen as an integral part of a more comprehensive community decision-making process. It was increasingly connected to issues such as environmental stewardship, community planning, and sustainable development (Britton, 1999). More analysis was devoted to emergency management as a necessary and vital component in broader community planning and development activities (Beatley, 1995; Mileti, 1999). The linkage of hazard mitigation, as a newly enhanced emphasis or priority in the emergency management cycle, to the broader task of developing sustainable communities placed emergency management at the heart of community development and planning (Schneider, 2002). A growing consensus seemed to emerge and suggest that the task-oriented, technical, and disaster-specific orientation of the old emergency management must be replaced with a broader and more strategic framework. This framework, in turn, seemed to many a likely candidate for the foundation of emergency management as a profession.

The new framework for emergency management that began to emerge suggested that its operational and technical capacities needed to be linked to the policy setting and stakeholder support bases of the communities it serves. Emergency management organizations were encouraged to see themselves as part of the dynamic political and social settings in which they worked and as having dynamic qualities of their own that would enable them to change, adapt, see challenges, identify opportunities, and create a broader role for themselves in the broader process of community planning and development. Most of this movement toward what might be called a new emergency management and an orientation connecting it to the broader work of community development can, in the judgment of many, be directly linked to the national emphasis during the 1990s on hazard mitigation as a priority and its linkage to the concept of sustainability.

THE NEW EMERGENCY MANAGEMENT

By the end of the twentieth century, emergency managers following the lead of FEMA and their state Divisions of Emergency Management were responsible for implementing an all hazards (natural, industrial, and all others) and all phases (preparedness, response, recovery, mitigation) approach to emergency management within their local jurisdictions. This became the model for emergency management in the United States. The rationale for this was ingrained within the trade. Whether the levees in New Orleans, for example, were breached by a hurricane or by a terrorist bomb, the response capacities required would be the same. The idea was to develop capacities and knowledge that were applicable to all disasters whatever their origin. Likewise, the emergency management cycle logically was understood to include the necessity for the integration of efforts (local, state, and national) across all disaster phases. Hence, the emergency management practitioner came to understand that all hazards within a jurisdiction must be considered as part of a thorough risk assessment, and they must be prioritized on the basis of their potential impact and likelihood of occurrence. Likewise, all disaster phases upon which the comprehensive emergency management model is based must be coordinated and managed as part of an integrated intergovernmental emergency management function. But another important development, a new emphasis on disaster or hazard mitigation, may have been even more important in redefining the nature and scope of the work in the field.

During the 1990s, it became an increasingly common premise in the field of public administration that the aim of managerial work in the public sector was to create public value (Moore, 1995). This was to suggest that public managers, including emergency managers, utilize scarce or limited public resources that have value in alternative uses. The challenge was said to be maximizing the public value attained through the expenditure of these resources. Managerial success in general came to be redefined as proactively initiating and reshaping public enterprises in ways that would increase their value to the publics they were intended to serve (Moore, 1995). Around this same time, some scholars began to talk about the concept of sustainability and its linkage to emergency management.

One need not delve too deeply into the public value discussion to see that the linkage of emergency management to the broader task of sustainable community development holds the potential for broadening the reach of emergency management considerably. It holds the potential for enhancing the public value of emergency management to the community it serves. It also holds the potential to redefine the purpose of emergency management and transform a trade into a profession. Such a linkage, as we shall see, is a challenge to recast emergency management as a participant, or even a leader, in the broad nexus of institutional, public, and private actors who influence the process of community planning and development. The linkage of disaster or hazard mitigation to sustainability would be the key to this transformation.

Sustainability to the emergency manager typically means that a lo-cality can withstand and overcome any damage (property damage, lost economic opportunity, etc.) caused by a disaster without significant outside assistance (Mileti, 1999). Hazard mitigation is the specific emer-gency management function that ties it most directly into the sustain-ability orbit. Mitigation is the assessment and management of hazard risks and community vulnerabilities in an effort to reduce the impact of disasters or, if possible, the likelihood of their occurrence in the first place. Fostering sustainability in the face of extreme hazard risks and events, natural or human-made, is a prominent theme in the current emergency management literature. An increasing number of federal and state initiatives during the 1990s encouraged and supported emer-gency managers, as they prepared to face and respond to specific and recurring disaster risks and vulnerabilities that confronted their com-munities, to think and act in terms of hazard mitigation.

The rationale for hazard mitigation begins with the realization that most disasters are expected. They stem from predictable interactions between the physical environment and the demographic and con-structed environments of the communities that experience them. Based on this realization, hazard mitigation that takes the form of advanced action to eliminate or reduce the risks and costs associated with disas-ters is both possible and practical. While Chapter 2 will entail a more detailed assessment of mitigation and its successes and shortcomings as an impetus to the professionalization of emergency management, we must note here that what one might call the "new emergency manage-ment" began with the emphasis on it in the 1990s.

In light of historic and rising costs associated with natural disasters in the United States over the 1980s and 1990s, it became, one might say, accepted wisdom within emergency management circles that a preem-inent objective of emergency management must be to mitigate hazards in a sustainable way to stop the trend of increasing and catastrophic losses from natural disasters. With the passage of the 2000 extension of the Robert T. Stafford Disaster Relief and Emergency Assistance Act, this accepted wisdom was embodied in federal law. In order to reduce the impact of recurring natural disasters such as hurricanes, floods, and earthquakes on human life and the constructed human environment, advanced planning to mitigate or manage the risks and vulnerabilities associated with them were among the things promoted by this legisla-tion.

A new emergency management may be said to have begun with the new focus on disaster mitigation. Since the beginning of the 1990s, emergency managers have become more intimately conversant with the concept of disaster mitigation. Federal grants and state funding have supported this development. FEMA and state-sponsored training emphasized it to a large degree. Structural mitigation became a popular theme. This included the strengthening of buildings and infrastructure exposed to hazard risks by a variety of soon to be well-known and discussed means (building codes, improved engineering designs, construction technologies, etc.). The purpose of structural mitigation is, of course, to increase resilience and damage resistance. Also much in vogue, as a topic for consideration if not always an accepted guide to action, was the concept of nonstructural mitigation. Nonstructural mitigation involves, for example, directing new development away from high-risk locations through land use plans and regulations, relocating existing developments that have sustained damage to safer locations, maintaining protective features of the natural environment that may absorb or reduce disaster impacts, and focusing on the sustainable economic and social well-being of the impacted communities. This new emphasis on disaster mitigation brought emergency management (potentially at least) into the center of a task it could not have previously imagined would be included in its more typical technical and narrow purview, the planning and implementing of sustainable community development.

The linkage of emergency management to the broader task of sustainable community development makes a great deal of sense. That does not mean the linkage is either inevitable or easy. But a consideration of the general rationale, together with the practical experience emergency managers have obtained in hazard mitigation over the past two decades, makes this linkage something worth exploring as the foundation for an emergency management profession.

Planning for sustainability, or sustainable development, is a concept originally associated with environmental policy. It has been broadened to include all community planning, including planning for economic development. It links concerns for social, economic, and environmental well-being in a coordinated process aimed at meeting present needs while preserving the ability of future generations to meet their needs. Emergency management has been increasingly linked to this broader task of sustainable development (Beatley, 1995; Geis and Kutzmark,

1995), and hazard mitigation has been a primary vehicle for that linkage (Mileti, 1999; Schneider, 2002). Perhaps of greater importance was the growing awareness that this new emphasis on reducing the vulnerability of communities to natural and human-made disasters could not be pursued without connecting it to all other goals associated with sustainable communities such as reducing poverty, providing jobs, promoting a strong economy, and generally improving people's living conditions (FEMA, 2000).

The achievement of sustainable development, as a public value, inevitably requires responsible choices for determining where and how development should proceed. It requires, from the emergency management perspective we might add and insist, an evaluation by each locality of its environmental resources and hazard risk potentials. It requires an analysis of the vulnerabilities they may have in the face of the identified risk potentials with the goal or purpose of making a series of choices that will positively impact (sustain) the environmental, economic, social, and physical well-being of the community. These choices include the identification of future losses that a community can or is willing to bear. But all public choices relating to these matters must adhere to the value of sustainability, however that may be defined in the context of the broader community planning and development process.

All emergency managers today know, by virtue of their experience and the federal and state mandates that define their work, that communities must address the interdependent causes of natural and human-made disasters and come to some decision about which potential risks and losses that may be anticipated are acceptable, which are unacceptable, and what specific actions are necessary to maintain the social, economic, and political stability necessary for the community to flourish. However, they seldom perceive this in the context of a broader role for emergency management in community planning. But consider the obvious connection between the two. For example, if a community is seeking to promote sustainability in the face of serious earthquake risks, structural mitigation alone is insufficient. Much more is required than building codes and the like. Sustainability also requires a linkage of policies on building codes to policies on housing density, to policies on urban transit, to policies on social equality, to policies on environmental quality, to policies on economic development, and so on. In other words, all policies are linked together by the concept of sustainability. This includes emergency management policy, and it brings the emergency management function, ideally, to the table as a necessary participant in community planning.

The goal of building sustainable communities must involve, and as a critical component, a broader role for and a richer involvement of the emergency management function. The logic of hazard mitigation suggests that a part of ensuring the economic, political, and social development of a community is a full awareness of hazard risks and vulnerabilities and a plan to mitigate them. Community planning and development must include anticipation of and solutions to the identifiable risks associated with these potential hazards. But, to the extent that emergency management is unprepared as a profession to assert its relevance to the broader life of the community, to the extent that it remains disaster specific, narrow, reactive, and technical in its orientation, the effectiveness and relevance of a new emergency management will be restricted even if there is a greater and growing awareness of its connection to broader issues and concerns. The new and wider context of emergency management requires a new and more broadly engaged emergency management professional.

THE NEW EMERGENCY MANAGER

As the discussion of a broader role for emergency management continued throughout the 1990s and into the new century, one in which the mitigation function tied it to sustainable community development, a discussion inevitably followed regarding the need for a new emergency manager to fill that role. By 2007, the work of a study group of academics and practitioners published by the International Association of Emergency Management (IAEM) culminated in a statement of the principles of emergency management. These principles presented a reasonable picture of what the new emergency manager ideally looks like (IAEM, 2007). To some extent, they reflected a new self-image. Emergency management was seen now not as merely a technical function but as a managerial one. The primary managerial function was to create a framework within which communities reduce their vulnerabilities to disasters. Reducing vulnerabilities to disasters became the essential task. Emergency management was articulated as the coordinating and integrating of all activities necessary to promote safer, less vulnerable, and more sustainable communities (IAEM, 2007). It is interesting to examine the eight principles identified by this group and to note the persistent theme of improving a community's ability to build, sustain, and improve the capacities needed to mitigate, prepare for, respond to, and recover from disasters.

IAEM's PRINCIPLES OF EMERGENCY MANAGEMENT

Emergency Managers Must Be:

1. Comprehensive – emergency managers consider and take into account all hazards, all phases, all stakeholders, and impacts relevant to disasters.
2. Progressive – emergency managers anticipate future disasters and take preventive and preparatory measures *to build disaster-resistant and disaster-resilient communities.*
3. Risk Driven – emergency managers use sound *risk management principles* (hazard identification, risk analysis, and impact analysis) in assigning priorities and resources.
4. Integrated – emergency managers *ensure unity of effort* among all levels of government and all elements of a community.
5. Collaborative – emergency managers *create and sustain broad and sincere relationships* among individuals and organizations to encourage trust, advocate a team atmosphere, build consensus, and facilitate communication.
6. Coordinated – emergency managers synchronize the activities of all relevant stakeholders to achieve a common purpose.
7. Flexible – emergency managers use creative and innovative approaches in *solving disaster challenges.*
8. Professional – emergency managers value a science- and knowledge-based approach on education, training, experience, ethical practice, public stewardship, and continuous improvement. (IAEM, 2007, italics added)

Aside from tights, a cape, and an "S" emblazoned on their chests, what do emergency managers need to live up to these principles? Certainly implied is a specialized body of knowledge that includes a study of historical disasters and an in-depth awareness of best practices in mitigation, response, and recovery. Likewise implied is a rich knowledge of social science and policy literature relating to disaster issues large and small. Training in the management of public organizations, organizational leadership, government, and politics comes to mind. Standards and guidelines for emergency management practice, an agreed-on definition of the emergency management function, an agreed-on code of professional ethics, and the connecting of the emergency management function to larger policy issues that shape its working environment and its chances for success (not to mention the importance of emergency management to them) and some means of certification and recognition of it as a professional managerial role. This is some list of needs, incomplete to be sure, but logical steps would seem to advance this vision of the new emergency manager.

If, as suggested in this analysis, emergency management is to become a critical part of the process of sustainable community development, its practitioners must first come to see their work in a new and broader context. Emergency managers must see themselves as participating with all political and social institutions in a coordinated effort. The primary focus must be on the building of sustainable communities as the fundamental public value to be served by the emergency management function. But the question remains how, in the performing of their specific tasks, can emergency managers organize their work to serve this public value?

As a first step, emergency managers must be trained and prepared to articulate and develop a role for themselves as participants in the local consensus-building effort in their community and to perceive themselves as working on a common agenda with other community institutions and leaders. All relevant public and private stakeholders, as defined in the context of sustainable development, must be brought into the emergency management planning process. Emergency managers, in turn, must be brought in as stakeholders and valued participants to the network of community leaders and policymakers involved in community planning and development activities.

A second step, to be accomplished as the emergency management function is integrated into the process of community planning, is the definition of the technical components in each phase of the function (risk assessment, mitigation, preparedness, response, and recovery) as part of a holistic system. This entails, as the most fundamental ingredient, the integration and consistency of all technical components with integrated policies and programs related to disaster mitigation as a necessary part of sustainability and to sustainability as the primary goal within the community. Hazard or disaster mitigation, in essence, must be the essential task that ties emergency management to the value of sustainability and defines a role for it in the context of community planning and development. Therefore, it must be elevated to the level of first or primary responsibility for the emergency manager. It also must be elevated to the level of an essential or a necessary component in all community planning and development activity.

A third or final step necessary for the new emergency management and the new emergency manager to succeed as a component in broader community development is the linkage of all public policies within the community to the concept of sustainability. All policies needed to promote the social, economic, and political stability necessary for a community to flourish, including emergency management policies, must be linked or integrated in the process of community planning. The end product of emergency management must be understood as fundamentally connected to all facets of community life in a coordinated effort with all relevant actors, public and private, to promote sustainability. This means that, in addition to the technical skills that relate to each disaster phase in the emergency management cycle, emergency managers must bring knowledge and a perspective to the table that are relevant to the broader task of sustainable community development.

To accomplish the three steps briefly outlined here, emergency managers need to move from the traditional, and still tempting as well as somewhat more comfortable, tendency to be reactive and disaster-specific. They must broaden their orientation beyond efficient disaster response and recovery operations. They must not be subject to the impulse to be reactive but, rather, strive to be assertive. Everything we have discussed requires that emergency managers be more proactive by emphasizing hazard mitigation. This ultimately means they must work to become networked in partnerships that involve all community leaders associated with the concept of sustainable development. They must become key public actors in the context of a broader involvement in community planning. To make the necessary and persuasive case to community leaders, to build networks of support groups and stakeholders, and to establish the strategic linkages with other community leaders and institutions necessary to bring about this transformation, technical skill alone is insufficient. The training and education of emergency managers needs to be refocused on the skills relevant for a more strategic emergency management.

Increasingly, it would seem, advanced educational training at the undergraduate and perhaps even graduate levels may be desirable for all emergency managers. Many practicing emergency managers would most likely disagree with this statement, but the argument for this is gaining strength as the demands of the trade are expanding and the calls for the professionalization of it are on the rise. The sort of training associated with public administration, including advanced training in leadership, organizational behavior, strategic planning, analytical methods, and public policy, has never been more urgently needed one could argue. The challenge of articulating a broader role for emergency management, its vital linkage to the building of sustainable communities, and its need to emphasize mitigation all suggest that a more proactive professional is needed. The vital tasks of networking and building relationships within the community of decision makers, the ability to recognize the opportunities for successful hazard mitigation in the broader community task of sustainable development, and the need for strategic thinking and leadership all demand that the education of the emergency management practitioner take on a new priority and that it represent a broader range of managerial and intellectual competencies than those typically associated with the important technical skills in the field.

Finally, it may also be logical and necessary that the professional training of all public management professionals should include a basic foundation in emergency management. Graduate and undergraduate programs alike should provide more training that reflects the linkage among hazard mitigation, community planning, and sustainable development. This does not mean that all public administrators, for example, should be cross-trained as emergency managers, but rather that emergency management should be a required component of their professional education. This should include at a minimum a focus on the value of mitigating hazards in a sustainable way as a critical and necessary component to community planning and development generally. Such training will broaden the understanding that the assessment of hazard potentials and the mitigation against their potential impact is connected to the making of a series of choices that impact the economic, physical, and social well-being of the community.

It is interesting to see the actual training that emergency managers are receiving or have available to them in the context of the perhaps overly idealistic portrait we have just suggested. One of the most widely known programs is the Certified Emergency Manager program (CEM) offered by the IAEM. The IAEM certificate program is not a requirement per se in the field, but it is a method of individual certification that practitioners find useful, and it has growing credibility in the field. It is a program for individuals already in service, and it represents an effort by the IAEM to offer some standard form of certification that might eventually lead to the professionalization of emergency management. An examination of what this certificate program is and the qualifications for entry into it may be instructive.

International Association of Emergency Managers CEM Program

What Is a Certified Emergency Manager?

Here are just a few of the reasons that many employers now list the CEM® as a job requirement when posting open positions for emergency managers:

- A Certified Emergency Manager (CEM) has the knowledge, skills, and ability to effectively manage a comprehensive emergency management program.
- A CEM has a working knowledge of all the basic tenets of emergency management, including mitigation, preparedness, response, and recovery.
- A CEM has experience and knowledge of interagency and community-wide participation in planning, coordination, and management functions designed to improve emergency management capabilities.
- A CEM can effectively accomplish the goals and objectives of any emergency management program in all environments with little or no additional training orientation.

Why Become a Certified Emergency Manager?

There are many reasons that emergency managers decide to pursue certification as a Certified Emergency Manager. Here are some of the benefits:

- Receive recognition of professional competence.
- Join an established network of credentialed professionals.
- Take advantage of enhanced career opportunities.
- Gain access to career development counseling.
- Obtain formal recognition of educational activities.

Requirements for the Certified Emergency Manager Program:

- Emergency management experience. Three years by date of application. Comprehensive experience must include participation in a full-scale exercise or actual disaster.
- Education. A four-year-baccalaureate degree in any subject area.
- Training. 100 contact hours in emergency management training and 100 hours in general management training. Note: No more than 25% of hours can be in any one topic.
- Contributions to the profession. Six separate contributions in areas such as professional membership, speaking, publishing articles, serving on volunteer boards or committees, and other areas beyond the scope of the emergency management job requirements.
- Comprehensive emergency management essay. Real-life scenarios are provided, and response must demonstrate knowledge, skills, and abilities as listed in the essay instructions.
- Multiple-choice examination. Candidates sit for the 100-question exam after their initial application and the other requirements are satisfied. The exam is a maximum of two (2) hours. A pamphlet is available further describing format and sources.
- Three references (including a reference from the candidate's current supervisor).

Note: A baccalaureate in emergency management reduces the experience requirement to 2 years and waives EM training if the degree was earned recently (IAEM, 2012).

The CEM is one effort to credential emergency managers and advance the trade toward a professional status. Its generic nature (all tenets of emergency management, including mitigation, preparedness, response, and recovery), its purpose of enhancing the knowledge and skills necessary to manage a comprehensive emergency management program, and its articulated benefits (recognition of professional competence, formal recognition of educational activities, career advancement, etc.) speak of some of the very things emergency management needs as it moves toward professional status. The same can be said with respect to emergency management training in higher education. With respect to formal educational and degree opportunities, these are multiplying and gaining in popularity. At the very least, the number of undergraduate and graduate programs suggests the recognition of a need for some sort of advanced training.

According to the FEMA higher education program, by the middle of the first decade of the 21st century, there were 141 college-level emergency management programs: 46 were certificate based, minors, tracks, and so on; 35 were associate degrees; 20 were bachelor's degrees; 34 were master's-level programs; and 6 were doctoral-level programs (Blanchard, 2008). One important note to mention is that emergency management is not an academic discipline. It does not have a body of knowledge, theory, scholarship, and so on that provides a unique and coherent disciplinary framework for student instruction. Emergency management is thus of necessity interdisciplinary or multidisciplinary in nature. This is not a bad thing at all and is, in fact, most desirable. But lacking the attributes of a single, tight, identifiable, and structured discipline may contribute to a multitude of degree programs representing a multitude of variations on a yet to be more fully defined theme. Such programs that exist (all of them presumed excellent for purposes of this discussion) may have homes in a variety of academic disciplines including sociology, criminal justice, public administration, geography, engineering, and health sciences. The curriculum of each program, while having commonalities with others, is usually designed to fit the contours of whichever discipline houses it.

The existing emergency management programs in higher education cover a wide span of academic disciplines but often with a narrow scope. While designed to meet the perceived need for a new type of emergency manager with broader sets of managerial skills, they often still seem to resemble or lean toward the more specific and technical perspectives associated with the old emergency manager. Programs are designed to prepare students for all of the basics (i.e., disaster preparedness training, plans and procedures for natural disasters, technological disasters, and even terrorism, and the competencies required to navigate in federal, state, and local governmental settings), but the focus is frequently on first responders (i.e., fire, law enforcement, public health, offices of emergency services, and other specific response agencies). There is, of course, nothing wrong with all of this. But it does not really produce the new emergency manager so much as it makes the old one more "busy." It may add to the managerial kit and practical skills of the emergency manager, but something is missing. Its task and technical focus seems to overwhelm other themes and is limiting to some degree. It does not really, as we have said, broaden the understanding that the assessment of hazard potentials and the mitigation against their potential impact is connected to the making of a series of choices that impact the economic, physical, and social well-being of the community. It does not necessarily tie emergency management into the broader task of creating hazard resilient or sustainable communities as a first priority. Rather, it seems to build new layers onto the functional, technical, and managerial components of the job but not to lift the trade to the status of a profession. In addition to often seeming somewhat uninspiring, it does not provide what is most needed. It does not provide the new and wider context that emergency management requires for the creation of new and more broadly engaged emergency management professionals.

It may be said with more than a little justification that the idealistic portrait of the new emergency management and the new emergency manager that has been articulated herein is well beyond the worldview and experience of any practicing emergency manager. Emergency managers, whatever their differing backgrounds, educational levels, and experiences, for the most part share an outlook or a mindset that makes much of the discussion about the new emergency manager, however interesting, not reasonable to them. Their focus is on immediate and extreme events (hurricanes, floods, earthquakes, winter storms, etc.), and their concentration is on preparing for them. Their interest in hazard mitigation (beyond the meeting of federal requirements) is limited to the question of whether it will make things better during the next flood, storm, hurricane, and so on (Labadie, 2011). Sustainability is a nice concept, but it is ultimately seen as the responsibility of others. Their time frame is limited to narrow planning cycles (2–5 years), and their work is often a response to federal or state mandates.

A study of the King County, Washington Office of Emergency Management showed some interesting information about the actual allocation of its staff time (Stehr, 2007). Over a seven-year period (1999–2006), the total number of staff hours expended increased from 12,417 to 31,921. But this increase (155%) was due primarily to the expansion of the staff from six to seventeen over that time period. Interestingly and more to the point, only seven percent of the staff time was spent on mitigation, response, and recovery-related activities (2.6% on mitigation, 3.1% on response, 1.3% recovery). Response and recovery were low due to the fact that the response and recovery activities are widely distributed among local governmental jurisdictions within the county.

Mitigation likewise was decentralized and undertaken by the local units having the primary legal authority to impose planning and land use requirements. The largest allocations of time were for disaster preparedness (39.7%) and grant management (31.6%). Coordinating the planning activity of others in the incident response system, disaster preparedness, and grant management ate up most of the time (Stehr, 2007). One would expect that this is fairly typical. Even where specific communities may spend more time on mitigation, response, and recovery activities, it is likely the case that disaster preparedness, a short time horizon, an immediate focus for planning activities, an emphasis on extreme and more immediate events, and administrative concerns (grants and others) dominate all other things that compete for the practicing emergency manager's time. This may lead one to wonder what the prospects really are for the new emergency management and the new emergency manager we have been discussing.

THE FUTURE OF EMERGENCY MANAGEMENT

It has been said that emergency management over the past two decades has become more "management" and less "emergency" (Britton, 2001). Consider the themes that have emerged over the past two decades as the role of emergency management has evolved. These themes definitely indicate an expanding and changing role for it and its practitioners. But has this expanding and changing led to the creation of a profession?

The role of the emergency manager is now regarded as much more than an operational role. It is increasingly an administrative or a managerial one that requires new knowledge and skills (Moore, 2010). Emergency managers are coordinators of activities (preparedness, response, mitigation, and recovery) that require combining the efforts of multiple agencies, intergovernmental cooperation (local, state, federal), and participating in and the building of public and private networks for collaboration (Waugh, 2011). There is a growing awareness that emergency managers face new challenges imposed by environmental and economic issues that, while outside of their traditional and normal range of activities, they must increasingly take into account and be knowledgeable about (Lindell, Prater, and Perry, 2007). With the brief ascendency of disaster mitigation as a priority in the 1990s, emergency management was increasingly seen as a part of the broader goal of creating hazard-resilient and sustainable communities and an integral component in community planning and development generally. This, in turn, created the potential for mainstreaming emergency management (i.e., integrating its goals, objectives, and initiatives as essential components into overall community planning and development activities). All of these developments stimulated the discussion of emergency management as a profession, at least potentially, and spurred the ongoing efforts to improve the training and formal educational opportunities for practitioners. But has this led to the formation of an emergency management profession? It might be argued it has, but an examination of what is missing may lead to another conclusion.

For all of the progress made and all of the impressive discussion of the trending of emergency management toward professional status, much remains to be done to complete that transition. First, the issues of certification and accreditation are unresolved. The CEM program provided by the IAEMs (and endorsed by FEMA) is an attempt to provide some leadership in that area. But this certification, much respected and highly recommended, is not required. Likewise, although an explosion of emergency management programs in higher education has occurred, there are no degree requirements for entry into the trade. Indeed, many of the individuals hired or appointed to emergency management positions do not have either a four-year degree or the CEM. Frequently they are individuals with backgrounds in law enforcement or the fire department who have had some disaster-related responsibilities or experiences, but there is great variance with respect to formal training and educational backgrounds. Further, many hiring agencies and entities do not really know what they are hiring or how to evaluate the resumes of those who apply. They simply gravitate toward someone with a law enforcement or response agency background. The reasons for this may relate directly to what is missing from the profile of the trade and inhibit it from becoming a profession.

The absence of an agreed-on process for certification and the lack of formal credentialing results in the lack of any structured formation for a profession. Qualifications for entry into the trade and for advancement from entry to mid-level to advanced or senior level remain undefined. This lack of a hierarchical structure and the absence of recognized and agreed-on criteria for its autonomy in governing itself as a profession retard progress. But these concerns may be rather easily resolved once and if emergency management is able to move past the technical and specific functions of the trade, and all of the impressive bells and whistles that go with them, to a broader sense of self. The trade has matured and become more technically involved and managerial, clearly, but it is not yet a fully grown professional adult. The necessary ingredient is to decide what it wants to be when it grows up. Does it wish to remain a trade, however technical and complex, or does it truly want to be a profession? In all probability, the issues of certification, credentialing, hierarchical structure, and professional autonomy will never be fully resolved until emergency management decides what it actually is when all is said and done. This necessary ingredient (i.e., the defining of emergency management as a profession) will help to focus all of these issues more precisely.

Emergency management has always been subject to being event-driven or disaster-specific, with its primary focus on response and recovery operations with a narrow emphasis on its technical capabilities. Even the expanding managerial role for the emergency manager as a coordinator of all hazards and all phases within their jurisdiction, and managing the often complex intergovernmental partnerships with state and national entities, has not been enough to overcome that tendency by itself. The notion of a new emergency management, one that may be the foundation for a profession, gained its only real traction as the emphasis on hazard mitigation and its connection to the concept of sustainable development ascended for a brief time. During the 1990s, this linkage seemed to hold great potential. It argued that the technical components of emergency management must be seen as part of a holistic and more strategic system that connects the emergency manager to the broader concerns of community planning and development. This was said to require the integration and consistency of all technical components of the craft or trade with policies and programs related to disaster mitigation as a necessary and critical component in the broader process of building sustainable communities. Resident in this view was both the opportunity and need to broaden the definition of the emergency management function. This broadening or redefinition, in turn, would require a more broadly trained, strategic, and proactive emergency management professional.

The notion that emergency management was a vital component in the building of sustainable communities with a unique knowledge, method, skill set, and holistic perspective that gave it an authoritative stature and an important role to play that required it to have a seat at the table in all community planning, policy, and development activities never fully launched. The reasons for this will be fully explored in Chapter 2. But this, or something much like it, is exactly what emergency management needs most to advance toward becoming a profession. It needs to carve out its niche in the broader context of community development. It needs to be recognized as having a body of knowledge and expertise that connects it to the community comprehensively and essentially. It needs to be a recognized authority with something to contribute that is unique to it and absolutely necessary for the broader task of sustainable community development. It needs to be valued and integrated into community life for more than its technical contributions. It must be understood as vitally connected to the broader concerns of human and social life and an integral part of sustainable community development. This includes the recognition of the usefulness and necessity of including its professional perspective on an array of issues not typically thought of when focused only on its narrower functions and technical skills. While individual practitioners may of necessity work in one or more specialized technical areas, the profession as a whole (should it ever come to exist) must have a voice that speaks more comprehensively and effectively, providing its special knowledge and methods to the assessment of all issues where its perspective has the potential to make a useful and necessary contribution.

Establishing the linkage of emergency management more emphatically to the task of building resilient and sustainable communities, it will be suggested, is an essential and necessary step for the development of an emergency management profession (Chapter 2). This connection, to the degree that it defines emergency management, also must come to serve as a foundation for the development of a code of ethics for the profession (Chapter 3). Of even greater importance, this connection ties emergency management as a participant in the broader range of important and critical issues where its perspective may contribute to the analysis of a wide array of problems impacting the specific prospects for sustainability and the future of humanity more generally (Chapters 4 and 5).

The transformation of emergency management from trade to profession, from occupation to vocation, requires that more attention be paid to the mental or intellectual foundations than to the manual work or specific technical functions that come with the job. These latter have evolved nicely and will, as they must of course, continue to do so. But the foundations have not been properly laid to guide an evolution of an ever-more complex trade to a genuine professional status. Until they are, emergency managers will forever be telling elected officials that their job is to tell them things they don't want to hear, asking them to spend money they don't have, for something they don't believe will ever happen.

REFERENCES

Beatley, T. (1995). Planning and Sustainability: A New (Improved?) Paradigm. *Journal of Planning Literature, 9*(4), 383–395.

Blanchard, W. B. (2008). "FEMA Higher Education Project Presentation, Slide: Emergency Management Collegiate Programs in 2006." July 1, 2008. Accessed at http://www.training.fema.gov/emiweb/edu/.

Britton, N. R. (1999). Wither Emergency Management? *International Journal of Mass Emergencies and Disasters, 17*(2), 223–235.

Britton, N. R. (2001). A New Emergency Management for a New Millennium? *Australian Journal of Emergency Management, 16*(4), 44–54.

Clement, K. E. (2011). The Essential of Emergency Management and Homeland Security Graduate Education Programs: Design, Development, and Future. *Journal of Homeland Security and Emergency Management, 8*(2), article 12.

Crews, D. T., CEM. (2001). The Case for Emergency Management as a Profession. *Australian Journal of Emergency Management, 16*(2), 2–3.

Cwiak, C. (2011). Framing the Future: What Should Emergency Management Graduates Know? *Journal of Homeland Security and Emergency Management, 8*(2), article 14.

FEMA. http://www.fema.gov/about/

FEMA. (2000). *Planning for Sustainability: The Link Between Hazard Mitigation and Livability.* Washington, DC: FEMA.

Geis, D., & Kutzmark, T. (1995, August). Developing Sustainable Communities: The Future is Now. *Public Management,* 4–13.

Grant, N. K. (1996). Emergency Management Training and Education for Public Administrators. In R. T. Sylves & W. L. Waugh (Eds.). *Disaster Management in the U.S. and Canada* (pp. 313–325). Chicago, IL: Charles T. Thomas.

Haddow, G., & Bullock, J. (2005). *The Future of Emergency Management.* Washington, DC: Institute for Crisis, Disaster and Risk and Management, George Washington University.

IAEM. (2007). Principles of Emergency Management. Accessed at http://www.iaem.com/EMPrinciples/documents/POEMFlyerFINAL2007.pdf.

IAEM. (2012). CEM FAQ's. March 6, 2012. Accessed at http://www.iaem.com/certification/generalinfo/cem.htm.

Kreps, G. A. (1991). Organizing for Emergency Management. In T. E. Drabek & G. J. Hoetmer (Eds.), *Emergency Management Principles and Practices for Local Government* (pp. 161–200). Washington, DC: International City Managers Association.

Labadie, J. R. (2011). Emergency Managers Confront Climate Change. *Sustainability, 3*(8), 1250–1264.

Lindell, M. K., Prater, C., & Perry, R. W. (2007). *Chapter 14: Professional Accountability. Emergency Management.* Hoboken, NJ: John Wiley & Sons.

Mileti, D. S. (1999). *Disasters by Design: A Reassessment of Natural Hazards in the United States.* Washington, DC: Joseph Henry Press, Environmental Studies.

Moore, A., CEM. (2010). The Future of Emergency Management: What Shall We Be? Accessed at http://www.docstoc.com/docs/44743003/THE-FUTURE-OF-EMERGENCY MANAGEMENT-WHAT-SHALL-WE-BE.

Moore, M. H. (1995). *Creating Public Value: Strategic Management in Government.* Cambridge, MA: Harvard University Press.

Oyola-Yemaiel, A., & Wilson, J. (2005). Three Essential Strategies for Emergency Management Professionalization in the U.S. *International Journal of Mass Emergencies and Disasters, 23*(1), 77–84.

Perry, R. W., & Mushkatel, A. H. (1984). *Disaster Management: Warning, Response, and Community Relocation.* Westport, CT: Quorum.

Schneider, R. O. (2002). Hazard Mitigation and Sustainable Community Development. *Disaster Prevention and Management, 11*(2), 141–147.

Stehr, S. D. (2007). The Changing Roles and Responsibilities of the Local Emergency Manager: An Empirical Study. *International Journal of Mass Emergencies and Disasters, 25*(1), 27–55.

Waugh, W. L. (2011). Emergency and Crisis Management: Practice, Theory, and Profession. In D. C. Menzel & H. L. White (Eds.), *The State of Public Administration: Issues, Challenges, and Opportunities* (pp. 204–217). Armonk, NY: M. E. Sharpe.

Whitaker, B. (2007, September 9). Ready for Anything (That's Their Job). *The New York Times.* Accessed at http://www.nytimes.com/2007/09/09/jobs/09starts.html.

Wolensky, R. P., & Wolensky, K. C. (1990). Local Government's Problems with Disaster Management: A Literature Review and Structural Analysis. *Policy Studies Review, 8,* 703–725.

Wright, R. P., & Rossi, P. H. (1981). *Social Sciences and Natural Hazards.* Cambridge, MA: Abt Books.

Chapter 2

MITIGATION AND SUSTAINABILITY: IT TAKES A VILLAGE

We shall require a substantially new manner of thinking if mankind is to survive.

— Albert Einstein

INTRODUCTION

In general terms, sustainability is the effective use of resources (natural, human, and technological) to meet today's needs while ensuring that these resources are available to meet future needs. Thus, sustainable development is the meeting of today's needs by communities without compromising the ability of future generations to meet their needs. Hazard or disaster mitigation, in this context, is part of fostering sustainable communities in the face of the risks and vulnerabilities that potential and extreme hazardous events inevitably impose on them. This is to say, as noted in Chapter 1, hazard mitigation is the emergency management function that places it at the center of sustainable community development and decision making.

The linkage of hazard mitigation to the concept of sustainable community development was a popular theme of the 1990s. In the United States, for example, as a response to the rising costs associated with natural disasters, it became a matter of national policy to regard the mitigation of hazards in a sustainable way as a priority. Among American scholars, hazard mitigation was increasingly connected to broader issues such as environmental stewardship, community planning, and sustainable development (Britton, 1999). The fostering of community sustainability in the face of disaster events (natural, technological, and all other) was viewed as integrally connected to the broader process of community planning and development (Beatley, 1995; Mileti, 1999; Schneider, 2002).

As the linkage of hazard mitigation to sustainable development became a popular theme, the early efforts by communities to implement sustainable hazard mitigation in practice were sporadic at best (Mileti, 1999). Likewise, the dedication of policymakers to mitigate hazard impacts before disasters struck and their commitment to plan and implement mitigation programs remained very much in question (Godschalk, Berke, Brower, and Kaiser, 1999). Nevertheless, the awareness that development decisions must take into account the forces that expose us to natural hazards (floods, earthquakes, hurricanes, erosion, wildfires, tornadoes, etc.), human-made hazards (pollution, toxic wastes, chemical spills, industrial accidents, etc.), and a variety of other hazard or disaster threats enhanced the importance of a broader reach for sustainability and the place of hazard mitigation under its umbrella. Before examining what that meant for emergency management, and the potential it held for the reformulation and redefinition of it, it may be useful to first elaborate on the linkage between hazard mitigation and sustainability.

SUSTAINABILITY AND THE LINKAGE TO MITIGATION

Defining sustainable development can be challenging to the extent that the concept of sustainable development may be a broad tent into which many groups and stakeholders can project their hopes and aspirations. The concept was first used by environmentalists who, in assessing problems associated with protecting threatened ecosystems, realized that these eco problems were connected to the lives of largely poor populations that dwelt in and around the sites of environmental concern (Kates, 2003). The interaction of the environment with development, the connection between the human condition (economic, health, nutrition, etc.) and the sustainability of communities, the importance of sustainable communities, and the linkage of the human condition to environmental sustainability led environmentalists to identify the concept of sustainable development as the vital center where everything intersects.

The United Nations Conference on Environment and Development in Rio de Janerio in 1992 signified the growing global acceptance of sustainable development as a goal. The promotion of broad-based development that is equitable, participatory, and environmentally sustainable, and that balances human and environmental needs with economic growth, became a shared concern in scientific and political communities throughout the world (Kusterer, Ruck, and Weaver, 1997). The fact that this goal seemed to be more urgently expressed by and on behalf of the developing world, and that it was of less immediate salience to many in the developed nations of the world, explains why it remains difficult to implement. Be that as it may, it is a more or less globally shared ambition to this day.

The linkage of sustainability to hazard mitigation, like its linkage to the human condition, is both logical and inherent. In general, whether in reference to environmental degradation, disease, or the impact of natural disasters, the majority of the threats faced by human communities are human-made. They stem from humanity's overwhelming success at dominating and transforming the world around it. They are connected to decisions that humanity makes about development, including and especially those decisions related to the constructed environment and economic development. Whatever the source of the threat, responding to a disaster or crisis once it unfolds is typically easier than recognizing disaster or crisis in the making and preventing or minimizing it. Reacting to disasters is also much more expensive and, in the long run, an unsustainable approach. Sustainable development requires, at a minimum, the development and use of knowledge relevant to building communities that can anticipate and prevent or minimize the impacts and costs of predictable disasters or crises. It encourages using knowledge to intelligently set goals, provides indicators and incentives, examines and evaluates alternatives, designs sustainable innovations, establishes effective institutions, encourages better decision making, and takes appropriate actions. This includes as a necessary component the anticipation and mitigation of disasters.

Sustainable development includes, to the degree it is indeed sustainable, the development of disaster-resilient communities. Disaster or hazard mitigation and sustainability are linked in a practical and necessary way. A brief elaboration of these two concepts in relation to each other demonstrates this connection clearly.

Disaster or hazard mitigation begins with the realization that most disasters are not unexpected. In fact, it has been suggested that all disasters may be considered disasters by human design (Mileti, 1999). That is, disasters are actually the predictable results of interactions between the physical or natural environment, the social, economic, and demographic characteristics of the human communities in that environment, and the specific features of the constructed environment (buildings, roads, bridges, and other features). To the degree they are in fact predictable, the magnitude of future disasters defined in terms of the risks to life, health, and property associated with them may be reduced with advanced planning. Effective mitigation, action taken to reduce such risks, can also substantially reduce the costs of disaster response and recovery. This is especially the case with respect to natural hazards such as hurricanes, floods, or earthquakes but applies to industrial, technological, and even economic disasters as well. With respect to each, forward thinking and advanced planning can reduce risks and contribute to the goal of sustainability.

With respect to natural hazards, two basic types of mitigation have proven to be most effective when employed: *structural mitigation* and *nonstructural mitigation* (also referred to as hard mitigation and soft mitigation, respectively). Structural (or hard) mitigation is the easiest to understand. It consists of the strengthening of buildings and infrastructures in the constructed environment exposed to natural hazards through a variety of means. These include building codes, improved engineering and design, improved construction technologies, and other practices that enable these structures to absorb or withstand exposure to natural hazards with minimal damage. The fundamental purpose of structural mitigation is to increase resilience and resistance to damage. As we shall see later in this chapter, such practices became common and were increasingly required by communities in the development of their local mitigation plans during the 1990s.

Nonstructural (or soft) mitigation is also important but for a variety of reasons can be more difficult to implement. Nonstructural mitigation includes directing development away from known hazard or high-risk locations through the adoption and implementation of land use plans and regulations. It also directs the redevelopment of communities that have suffered repeated damage in high-risk locations, relocating them to safer areas. It especially includes, to the degree that it is truly driven by the goal of sustainability, maintaining the protective features of the natural environment (i.e., sand dunes, forests, vegetated areas, etc.) that may absorb and thus reduce hazard impacts on densely populated and developed areas. Nonstructural mitigation also assesses and addresses the needs of at-risk populations. This includes housing and economic considerations as well as special needs populations like the elderly or the handicapped. Worldwide, the poorest populations are frequently exposed to the greatest risks and suffer the most devastation in relation to natural disasters. In a sense, the reduction of poverty could be considered a nonstructural or soft mitigation objective.

Nonstructural mitigation often encounters stiff resistance for a variety of reasons. Developers, for example, are not as interested in the preservation of protective features of the natural environment as they are in making money when it's time to build new beachfront resorts or hotels. Indeed the development practices in vulnerable coastal regions have, according to experts, enhanced what may be called the hurricane problem in the United States. The ever growing concentration of population and wealth in vulnerable coastal regions, combined with the destruction of protective natural features by the development decisions made to meet this demand, means that the United States is set up for rapidly escalating human and economic losses from hurricanes (Mooney, 2007). In other words, our vulnerabilities are increasing, in part, by human design. All too frequently, government policies do what is unsustainable (i.e., promote or subsidize risks) to appease more immediate (often short-sighted) economic needs or the demands of influential developers.

Whatever the impediments to hazard mitigation, and we will discuss some of these later in this chapter, it has long been known that effective mitigation reduces the cost of disaster response and recovery (Godschalk et al., 1999). Sustainability from an emergency management perspective means, at a minimum, that a community can tolerate and overcome the impacts of a disaster without significant outside assistance (Mileti, 1999). To achieve sustainability from an emergency management perspective also means that communities must choose where, when, and how development proceeds. It means that each community must evaluate its environmental resources, its hazard risks, and its vulnerabilities as it chooses the future losses it will risk, and their costs, or the necessary investment in policies and strategies that can minimize or reduce those costs and risks. Sustainability successes and failures are the result of a choice. It is a choice that, knowingly or unknowingly, all communities do make.

Planning for sustainable communities is connected to all community planning for social and economic development as well as to environmental well-being. This process balances the meeting of present needs with the ability of future generations to meet their needs. Community planning cannot meet the goals associated with sustainability if it does not ensure that economic and political decision makers operate with a full awareness of the risks to people and property vulnerable to natural, industrial, technological, or other hazards. This requires that community development, to be at all sustainable, must include anticipation of and solutions to the risks and vulnerabilities associated with predictable hazards. This, the hazard mitigation function, is what places emergency management at the center of community planning and development. Sustainability and hazard mitigation are thus linked at the hip. Before discussing what that might mean (has meant in fact) for emergency management, let us first turn to a brief discussion of the principles and basic techniques of sustainable hazard mitigation.

PRINCIPLES AND TECHNIQUES

Resistance and resilience in the face of predictable disasters are among the critical characteristics of a livable and sustainable community. As sustainable hazard mitigation has evolved and our understanding of its importance has matured, some basic and agreed-on first principles have emerged. These guiding principles, articulated best perhaps by Dennis Mileti in his book *Disasters by Design* (Mileti, 1999), are connected to the goal of sustainability and the understanding of it as the imperative that should animate all hazard mitigation activity. Let us briefly examine the six basic principles that have come to shape the goals of hazard mitigation in relation to sustainability (Mileti, 1999).

First, human activities in any community should be mindful of the need to maintain and, where possible, enhance environmental quality. This means that hazard mitigation efforts must be linked to natural resource management and protection. Not only is this important with respect to preserving natural features that may absorb hazard impacts or preventing a degradation of the surrounding environment, and thus exposing communities to greater risks of costly disaster impacts, it is also important with respect to the preservation and maintenance of resources (water comes to mind) that may be threatened and thus not available to future generations.

Second, sustainable communities must define and plan for the quality of life they want for themselves and for future generations. The quality of life, defined in terms of income, economic well-being, health, crime, pollution, recreation, disaster potentials, and associated risks and vulnerabilities, is a critical component in the planning, building, and maintaining of sustainable communities. Hazard mitigation planning must be seen as connected to all community planning relevant to the "quality of life" issue.

Third, sustainable communities foster local resiliency to and responsibilities for disaster. Each community is responsible for identifying, in a comprehensive fashion, its environmental resources and exposure to hazard risks. Hazard risks are here meant to be comprehensive and apply to natural, industrial, economic, and all other risks that may result from either natural causes or unsustainable human practices that may make a disaster inevitable. Reducing exposure to hazard risks, minimizing the potential for disaster impacts, is as important as preparing for and responding to disaster occurrence. It may even be considered more important if it results in implementing the necessary steps to withstand any potential hazard without significant loss or reduction in productivity or quality of life.

Fourth, sustainable communities are tied to healthy local economies. This principle suggests that local hazard mitigation is enhanced by a diversified economy that is not easily disrupted by disasters. Sustainable communities also ensure that disaster costs (present or future) are not shifted to other communities, at-risk populations within a community, the atmosphere, or future generations. This requires an effort to calculate risks accurately, distribute their costs fairly, and take into account the impact of economic decisions about growth, energy, employment, housing, etc. in relation to the goal of sustainability.

Fifth, a sustainable community preserves ecosystems and resources to ensure that today's advances do not pass on intolerable risks to future generations. Hazard mitigation cannot contribute to sustainability if it delays or postpones reasonable action only to pass on increased risk or higher costs to future generations. Likewise, hazard mitigation is rooted in an understanding that ecosystems do not expand and that the promulgation or continuation of unsustainable practices guarantees only the inevitable hitting of the wall in future generations.

The sixth and final principle of sustainable hazard mitigation emphasizes the need for a consensus-building approach that involves all people who have a stake in the outcome of hazard mitigation planning activities. This is literally everyone, from political and economic decision makers to residents. Broad participation in a community-wide dialogue, and the sense of ownership that may grow out of it, ensures that critical information will be generated, distributed, analyzed, and influential. There is a practical necessity that hazard mitigation be placed on the planning agenda of every public and private entity in the community. This can only contribute to a consensus-building effort that is critical to the shaping of a community-wide consensus that elevates sustainability, and sustainable mitigation practices, as the primary goal in the overall work of community planning and development. To be successful, hazard mitigation and the building of sustainable communities requires a culture that is participatory and all inclusive. Mitigation and sustainability must be a community-wide effort. Borrowing a phrase made popular during the 1990s in reference to the raising of children, *it takes a village to mitigate.*

Just as these six principles are today well known within the hazard mitigation community, so too are the basic techniques for sustainable hazard mitigation. A brief review of these principles emphasizes the relationship between minimizing risks and losses from hazards and the building of hazard-resilient and sustainable communities. The basic techniques include land use planning, building codes, insurance, warnings, engineering, and a variety of technological applications (Mileti, 1999).

Land use planning involves both the more efficient use of space and the reduction of hazard risks. It also means, in the context of our present discussion, that development and redevelopment decisions are connected to the preservation or restoration of natural protective features in the community. This is what it means when we say that disaster resilience is a component of local development policies. Local governments in the United States have utilized land use policy fairly extensively over the past two decades. The same may be said of building codes.

Building codes have been widely used to strengthen the constructed environment in the face of risks and vulnerabilities posed by natural disasters. Local governments in the United States have in fact enacted fairly comprehensive building codes to regulate new construction, and these codes increasingly reflect the priority of hazard mitigation. The model building codes that are available and utilized by state and local governments have become more widely used than custom-drafted codes. This is a positive reflection on the quality of these model codes. There has been, however, some concern raised regarding inconsistent or inadequate enforcement of codes. Inconsistent or inadequate enforcement efforts sometimes result in a reduction in the effectiveness of building codes as a mitigation technique. Nevertheless, the broad awareness that disaster-resilient construction is a critical component in sustainable community planning is widespread. Whatever the inconsistencies in enforcement, there seems to be universal agreement that it is a good thing if buildings do not fall down during earthquakes and businesses and homes are not destroyed by flooding, etc.

Insurance is not really a mitigation technique because it redistributes losses rather than reducing them. But insurance may minimize the disruption that often follows a disaster. It may also be a device that encourages people to adopt risk-reduction strategies if it creates incentives to mitigate by offering rate discounts or lower deductibles for doing so. U.S. flood control policies have emphasized flood insurance, required under certain circumstances, (100-year floodplains, for example), in an effort to establish an effective option to reduce government- and tax-payer-borne costs in disaster recovery and to improve the nation's floodplain management. The National Insurance Flood Program was designed to save the government money by using premiums paid by floodplain occupants to fund flood-related disaster assistance. Critics suggest, with some reason, that insurance may actually provide individual property and business owners with a disincentive to protect their property against flood damage. The assumption that losses will be covered may work against the initiation of costly mitigation strategies by property owners.

Improvements in the constructed environment made possible by improved engineering practices may contribute to the building of sustainability. In the building of new structures or the retrofitting of older ones, these practices may improve the resilience and hazard-resistant capacity of our infrastructure. But carefully engineered buildings and structures are not, in themselves, always a contribution to sustainable hazard mitigation. To the extent that improved engineering technology may encourage more expensive development in high-risk locations, it may lead to a false sense of security that actually increases rather than reduces hazard risks within a community. An overreliance on structural engineering to the exclusion of other measures may successfully postpone disaster damages for a time, but when structural engineering fails (as it often does), the damages may be greater than anticipated.

With respect to technology, the expanding variety of hazard-relevant applications is impressive. These applications can provide tremendous assistance to hazard mitigation efforts. Geographic Information Systems (GIS), for example, have many applications that may be used to estimate damages to infrastructure in various disaster scenarios. They can provide risk information to aid in community land use planning and in building design and construction, simulate disaster impacts, and aid in both developmental and environmental planning. Computer-mediated communication, remote-sensing, decision support systems, and risk analysis all have developed rapidly and are of great value in disaster preparedness and mitigation planning. While technologies have advanced, the ability of all communities to utilize and benefit from this advancement is unequal in that some communities simply cannot afford them or lack the expertise to use it effectively. Still, technological applications relevant for hazard mitigation are expanding, and their potential is seemingly unlimited.

The six principles and the six basic techniques just summarized demonstrate that the value of hazard mitigation and the techniques for sustainable hazard mitigation are well understood and, to varying degrees, being implemented. Increasingly clear is the notion that resilience and resistance to hazards is a characteristic of sustainable communities. By integrating hazard mitigation into the broader framework of sustainable development, communities have the ability to take effective action to eliminate or reduce the loss of life and property associated with disasters and to sustain economic vitality. Hazard mitigation is increasingly regarded as an essential component in the creation of safe and sustainable communities.

The principles and techniques of hazard mitigation ascended on the national agenda and transformed emergency management in the United States during the 1990s. To some extent, this transformation has endured. To some extent, it has been interrupted and, as such, remains incomplete. To understand this statement, let us turn to a discussion of that decade of promise and its aftermath.

THE MITIGATION DECADE

As the 1990s began, much of the emergency management literature discussed the obstacles that local governments and emergency managers had to overcome in order to establish and implement effective emergency management agencies. The obstacles included the diversity of hazards faced within communities; the technical complexity of regulatory, planning, and response efforts; the low salience of emergency management as a public issue; the historic and culturally ingrained resistance to regulation and planning in many local governments across the country; the lack of political and administrative advocacy for emergency management; the difficulty of communicating risks when the public and public officials doubt that proposed emergency management initiatives and programs will ever be needed; the jurisdictional confusion of a complex federal system and its division of authority among national, state, and local units; and an economic and political environment that was increasingly skeptical and inhospitable to government in general and new public programs in particular (Waugh, 1990). In the face of these obstacles, emergency management professionals tended to be narrow, disaster-specific, response-oriented, technical, and limited to their specific tasks. Not much was being done to overcome obstacles or transform emergency management into a more strategic position. But this was about to change.

The occurrence of what some called mega disasters (e.g., Hurricane Andrew in 1992 or the Midwest flooding in 1993) generated increased interest in and provided an impetus for emergency management in the United States. Of particular note at that time (i.e., the early 1990s), was the perceived and rapidly increasing recovery costs associated with recurring natural disasters. Floods, for example, were said to be the most serious disasters in the United States in terms of lives lost, people and communities impacted, property loss, and frequency of occurrence. Between 1966 and 1985, direct losses from floods grew from $2.2 billion per year to over $5 billion per year (Cigler, 1995). The Midwest flooding of 1993 was the most devastating flood in modern U.S. history up to that time, with economic damages near $20 billion (Kolva, 2002). It was becoming clear that natural disasters were in fact rapidly increasing in terms of both costs and frequency.

During the 1990s, the more serious disaster scenarios experienced in the United States saw disaster losses add up to more than $1 billion per week (FEMA, 1997). While there was some variation in estimates, annual losses from natural disasters averaged $10 billion per year between 1975 and 1994, and they were expected to increase. By way of comparison, Hurricane Katrina in 2005 saw disaster-related losses totaling $31.3 billion (in present-value dollars) on humanitarian aid, infrastructure repair, and funding for destroyed private property (Woolsey, 2008). In the five-year period between 1989 and 1994, natural disasters cost the U.S. treasury more than $34 billion (FEMA, 1997). Policymakers saw a need to act.

Until the 1990s, the amount of public money spent before a disaster (i.e., to mitigate or alleviate disaster vulnerability and reduce recovery costs) was a small fraction of that being spent after disasters. But the disaster-related experiences of the early 1990s and their escalating costs to the public treasury soon led to an emphasis on hazard mitigation as a response. A new perspective on natural disasters was emerging, and it emphasized that vulnerabilities are created by human actions and decisions (Anderson, 1995). Increasingly hazardous situations came to be seen in the context of the interface between extreme hazard events and vulnerable populations, and the vulnerability of populations came to be seen as the direct result of decisions and choices associated with community planning and development. It was clear to many that it was time to make better decisions. It was time to change the way we thought about both natural disasters and emergency management.

In 1993, the federal government and FEMA were indeed ready to change the way we think. James Lee Witt, President Clinton's FEMA director, was charged with the task of applying the Clinton administration's "reinventing government initiative" and, in the process, reinventing FEMA (Sylves, 1996). Witt was the first individual with an extensive background as an emergency manager appointed as director of FEMA. He was determined to reduce the costs of natural disasters in both economic and human terms with respect to lives lost or disrupted (Holdeman, 2005). Based on the blueprint provided in Osborne and Gaebler's famous book, *Reinventing Government* (Osborne and Gaebler, 1993), the Clinton administration compelled federal agencies to articulate strategies for applying the entrepreneurial spirit and tap into free market sources to improve efficiency and the service delivery capacity of the federal government. The result at FEMA under Witt was a significant transformation of emergency management in the United States.

The National Performance Review reports produced by FEMA during the "reinventing government" initiative refined its mission and defined new national priorities for emergency management. Attempting to give FEMA a clearer mission, Witt was determined to go beyond the more traditional emphasis on reactive or response-oriented activities. He emphasized a sharper focus on reducing risks as the first priority to be met (Sylves, 1996). He meant to reduce the repeated loss of property and lives every time a disaster struck (Holdeman, 2005). FEMA would now articulate a new emphasis on hazard mitigation.

Hazard mitigation soon became a stated and recognized national priority. It made good sense inasmuch as establishing mitigation as the number one priority (i.e., as the foundation for emergency management), would ideally decrease the costs and demands associated with disaster response and recovery. The plan ultimately produced by Witt called for the implementation of disaster mitigation planning by state and local governments. This included federal grant funds to assist states and communities in the development and implementation of mitigation strategies and a federal requirement that state and local governments show evidence of meeting mitigation standards as a condition for release by FEMA of federal disaster relief funding (Sylves, 1996). FEMA also promoted public–private sector collaborations in mitigation planning and worked to educate and include the public in communities across the country in a national mitigation dialogue. In the process, FEMA became a more independent and vital agency. The importance of FEMA's work and the degree to which hazard mitigation had become a national priority may be signified by the fact that, during the Clinton administration, FEMA was elevated to a cabinet-level status with the recognition that its responsibilities coordinated critical efforts that cut across both departmental and governmental lines. Direct access to the president, enhancing the FEMA director's authority and influence, reflected the degree to which emergency management in general and hazard mitigation in particular were national priorities of great importance for the Clinton administration.

Perhaps the most innovative and successful initiative produced to promote hazard mitigation, and to elevate its importance as a national priority, was the creation of a program called Project Impact. During the course of its lifetime, $20 to $25 billion was allocated to Project Impact to support local hazard mitigation as a national priority. The success stories associated with Project Impact demonstrated the value of hazard mitigation and contributed to a growing national consensus within the emergency management community about its importance.

Project Impact, formally initiated by FEMA in 1997, was designed to promote disaster-resilient communities. It promoted hazard mitigation across the country as a means of breaking the cycle of destruction and rebuilding that had become the norm in disaster scenarios. Disaster-resistant or hazard-resilient communities were said to be characterized not only by technical hazard mitigation efforts but also by efforts to promote sustainable economic development, efforts to protect and enhance the natural resources, and efforts to ensure a better quality of life for its citizens. Project Impact: Building Hazard Resistant Communities, as it was officially named, sought to build federal–state and public–private partnerships for hazard mitigation. This meant identifying risks, prioritizing needs, creating and implementing plans, and emphasizing private sector participation. FEMA, working with state offices of emergency management, designated local communities as Project Impact sites for project funding. Most of these projects targeted and recruited partners in business (e.g., insurance agencies, real estate firms, homebuilders, banks, etc.) and encouraged their support and practical involvement in mitigation efforts at the community level (Schwab, Eschelbach, and Brower, 2007).

By 1999, nearly 200 communities nationwide and more than 1,000 business partners were involved in Project Impact. It was generally estimated that for every dollar spent on hazard mitigation, more than two dollars were saved in disaster recovery and repair costs (FEMA, 1999). In Manhattan, Kansas, Project Impact investments paid off long after the program had ceased to exist. Project Impact grants for building "safe rooms" in the major areas of Manhattan that were subject to recurring tornado damages saved numerous lives when a half-mile-wide F3 tornado devastated the area in June 2008. Project Impact may have been discontinued in 2001, but in Dade County, Florida, its benefits were felt well after that time. All of Dade's major municipalities and major county departments, partnering with nine universities and colleges, ten hospitals, non-profits such as the Red Cross, and businesses including IBM, Wal-Mart, Macy's, Visa, UPS, American Airlines, etc., collaborated on the completion of $250 million worth of mitigation planning. The hurricanes of 2004 and 2005, a particularly active time in the American Southeast, were said to have produced far less damage than might otherwise have been expected as a direct result of this work.

Also, where flood mitigation had been done, there was no significant flooding damage (Holdeman and Patton, 2008).

Deerfield Beach, Florida, a coastal community of more than 66,000 people, was the first Project Impact community to partner with FEMA in 1997. Deerfield Beach, having been hit by seven major hurricanes throughout its history, was well acquainted with damages a natural disaster can invoke on a community. It was also clear to residents and businesses in Deerfield Beach that more hurricanes were an absolute certainty for their community. Despite this, city leaders were skeptical when FEMA first approached them with this new federal initiative called Project Impact. But they were eventually won over. Projects undertaken there included the retrofitting of the Deerfield High School (which also serves as a shelter during emergencies). Hurricane straps were added to the cafeteria, and wind shutters were placed on all of the school's windows. The Chamber of Commerce and City Hall were also retrofitted with hurricane-resistant windows and shutters. Soon, through public education efforts and volunteer programs, homeowners and businesses were retrofitting as well (FEMA, 1999). Before Project Impact was discontinued in 2001, FEMA had fostered partnerships among federal, state, and local emergency workers, along with local businesses, to prepare individual communities for natural disasters in all fifty states. In Seattle, Washington, project grants were used to retrofit schools, bridges, and houses at risk from earthquakes. In Pascagoula, Mississippi, the project funded the creation of a database of structures in the local flood plain (absolutely crucial information for preparing mitigation plans). In several eastern North Carolina communities, it helped fund and coordinate buyouts of houses in flood-prone areas. The list could go on and on.

In Seattle, Washington, a Project Impact pilot program initiated in 1998 led to the retrofitting of schools and more than 14,000 homes to resist earthquake damage. In 2001, a major earthquake rocked the Pacific Northwest, but its impact was significantly reduced as a direct result of the retrofitting. Ironically, on the same day that the earthquake hit, the Bush administration sent a budget to the Congress that eliminated the Project Impact program (Galvin, 2001). Before discussing the discontinuation of Project Impact and the status of mitigation in its aftermath, a discussion of the project's significance and impact is in order.

First, consider the meaning of this project and its goals for emergency management generally. In setting out to redefine FEMA's mission, to go beyond reactive response-related activities and the writing of checks after a disaster occurrence, a new agenda emerged. The determination to break the cycle of repeated property loss and increasing costs associated with each disaster led to a new focus and a clearly identified national priority. Reducing risks and building hazard-resilient communities became the first priority of FEMA and, by extension, emergency management nationally. This priority connected hazard mitigation as a critical component in the creation of sustainable communities. Resident in this priority was the foundation for both the new emergency management and the new emergency manager discussed in Chapter 1. It is worth repeating what was said there. The goal of building sustainable communities must involve, and as a critical component, a broader role for and a richer involvement of the emergency management function. The logic of hazard mitigation suggests that part of ensuring the economic, political, and social development of a community is a full awareness of hazard risks and vulnerabilities and a plan to mitigate them. Community planning and development must include anticipation of and solutions to the identifiable risks associated with these potential hazards.

It is likewise worth recalling what we said in Chapter 1 about the need for a new emergency manager. If emergency management is to become a critical part of the process of sustainable community development, its practitioners must first come to see their work in a new and broader context. Emergency managers must see themselves as participating with all political and social institutions in a coordinated effort. The primary focus must be on the building of sustainable communities as the fundamental public value to be served by the emergency management function. Clearly, these statements would seem to embody the new priority on hazard mitigation and the emphasis on building hazard-resilient communities that defined the 1990s at FEMA and across the nation.

The implementation of Project Impact initiatives across the nation also signified an effort to produce a significant change in our national thinking about the natural world and the threats it may present to the communities we build and rebuild. The objective of preventing disasters or reducing their impacts was promoted as an essential necessity. This would require a cultural change that is difficult to implement.

Building a culture of sustainability, mitigation, etc. is made difficult by the fact that the costs of mitigation are to be paid in the present but the benefits lie in the future. As noted in the previous chapter, most public attention is focused on disaster response and recovery. This is due to the fact that most people give little thought to natural or human-made disasters until they happen or are about to happen. The reluctance of policymakers and stakeholders to seriously anticipate disaster potentials and prepare for or mitigate and reduce their impacts had to be overcome. FEMA, during the Witt era, knew that the entire culture surrounding emergency management had to be changed.

The genius of Project Impact, if that word may be allowed, was in its commitment to building partnerships in local communities that involved public sector and private sector entities. First, hazard mitigation is best done locally as part of the planning and work necessary to create sustainable communities. This includes not just governmental and business entities, but home owners and individual citizens as well. As has been said and should be repeated often, mitigation takes a village. It is not a project for emergency managers alone, it is a community project. FEMA, understanding that it does indeed take a village, sought to promote the sort of comprehensive community-wide dialogue necessary to create a culture of sustainability and a habit of mitigation.

Creating a sustainability culture is no small undertaking. American culture generally is reluctant to take the long-range view. It is notoriously present-oriented and often oblivious to the fact that many short-term decisions and actions are unsustainable in the long run. This may seem harsh, but it is true. Americans sometimes seem to intentionally and recklessly enhance risks in pursuit of short-term goals or benefits and, as a consequence, create inevitable disasters in their wake. Nowhere is this more evident, perhaps, than with respect to the financial meltdown of 2008 that resulted in the worst financial crisis since the Great Depression. This disaster was born of decisions made by lenders and investors that enhanced risks to the breaking point, recklessly leveraged large amounts of debt for short-term gains, created a toxic and massive housing bubble that would inevitably burst, and pursued unsustainable practices for short-term profit and at the expense of long-term economic stability.

There are many examples of unsustainable practices that contributed to the collapse of the banking system. Subprime loans come to mind, for example, but many experts say the creation of credit-default swaps may have been the worst. These were little understood, unregulated, insurance-like contracts that were intended to protect against loan defaults. Toxic subprime housing loans were backed by trillions of dollars in credit swaps. The credit institutions that sold the swaps made short-term profits, but when home buyers began defaulting on loans (as they inevitably would) and the housing market crashed (as it inevitably would), financial institutions lacked the capital to make good on the guarantees, and investors who had purchased risky mortgage-backed securities were hung out to dry (Billiterri and Mattingly, 2010). All of this was inevitable and foreseeable. Unsustainable practices always hit the wall, and all bubbles eventually do burst. Whether it is an economic disaster or a natural disaster, this is how it plays out. Without attention to the future, the long view, and absent the identification, management, and mitigation of risks, sustainability is always the first casualty of short-sighted cultures.

During the 1990s, FEMA took the long view. The mitigation journey was begun in earnest, and communities across the country were aboard the mitigation train. In addition to the individually funded projects (some $20 to $25 million worth), Project Impact had initiated a national mitigation dialogue in localities around the nation. This is how cultural change begins. It also reoriented and redefined emergency management. In identifying hazard mitigation as a national priority, and in emphasizing its centrality to the emergency management function, FEMA made emergency management a more influential component in the process of building sustainable communities. It also sought to make hazard mitigation the work of all entities, public and private, within communities. This, in turn, led to the broader discussion of a new emergency management and the new emergency manager. The impetus to define emergency management beyond its technical functions, beyond the realm of first response agencies, and to define it in terms of not only a broader managerial context but in the context of its unique contribution to building sustainable communities, fueled the efforts to redefine emergency management. No longer defined exclusively by its technical functions and tools, emergency management was beginning the journey from trade to profession. But the journey is far from complete, as we suggested in Chapter 1, and it has been interrupted by events that followed the mitigation decade.

The end of Project Impact was not the end of hazard mitigation of course. It remains, to a lesser degree perhaps, one of the priorities of emergency management. But a discussion of Project Impact's aftermath, and the decade subsequent to its creation, does suggest that the initial progress has not been easy to sustain. Project Impact had its critics from the beginning, to be sure. Some said it did not really include all of the stakeholders that should be involved in community-based mitigation planning. It was criticized by some as being political with an emphasis on corporate style marketing campaigns. Others were at odds with what they perceived as its ideology and were suspicious of federal government initiatives in general and favored more local governmental autonomy. Inevitably, there were also debates about its cost (Schwab et al., 2007). It should thus be no surprise that, its successes and relative popularity notwithstanding, a partisan change in national leadership would contribute to its dismantling in favor of something else. But events also would play a role in altering the mitigation journey.

A DECADE ADRIFT

Perhaps the greatest flaw with FEMA's Project Impact was the fact that it relied on federal funding. There were no communities implementing Project Impact initiatives without federal grant spending. This is not unusual as local communities frequently prioritize an activity only if it has the potential to acquire some federal funding. This is also not to criticize FEMA for either its initiative or funding of local projects. But it does mean that as national priorities shift from one administration to the next, a new presidential administration with a different ideological orientation and a disdain for federally funded domestic projects may well discontinue even successful initiatives in the pursuit of other objectives.

In February 2001, Congress approved a proposal by the newly elected George W. Bush administration to terminate Project Impact. This did not mean that the federal government ceased its involvement in the promotion of hazard mitigation. The Disaster Mitigation Act of 2000, an amendment of the Stafford Act, remained on the books. It provided the legal basis for FEMA mitigation planning requirements for state, local, and Indian Tribal governments as a condition for receiving federal disaster assistance. The requirement of hazard mitigation planning as a prerequisite for receiving disaster assistance assured that the local governments had incentives to mitigate (Schwab et al., 2007). The Hazard Mitigation Grant Program (HMGP) authorized under Section 404 of the Stafford Act and administered by FEMA also remained intact. The HMGP enabled communities to obtain some federal funding for hazard mitigation efforts conducted in the aftermath or recovery phase of declared natural disasters. The logic of incorporating hazard mitigation into the rebuilding and recovery stage following a natural disaster makes perfect sense in that its objective is to make communities more disaster resilient in relation to natural hazards that are likely to reoccur on a relatively regular basis. Congress also instituted a new program under which pre-disaster mitigation grants could be awarded to communities on a competitive basis. Nevertheless, the termination of Project Impact indicated a shift in priorities.

The Bush administration and its new appointees showed little interest in the objectives pursued by FEMA under James Lee Witt. They argued that local communities would have greater flexibility and accomplish more without federal leadership on mitigation, and they questioned the value of the federal investments made in that effort (Holdeman, 2005). The new FEMA Director, Joseph Allbaugh, had different priorities as well. Allbaugh, who had been chief of staff for President Bush when he served as Texas governor and the national campaign manager for Bush-Cheney 2000 (and who had no background or experience in emergency management), would place his major emphasis on disaster response and recovery. Agreeing with the sentiment that Project Impact was of limited value, he sought more federal funding for first responder training and equipment purchases (Bergal et al., 2007). This marked a return to a more traditional response orientation and a focus on first responders.

In 2002, the Bush administration would make dramatic cuts in federal funding for hazard mitigation grants, asserting that the states would select more cost-effective and efficient programs if they paid more for them out of their own pocket, an incentive they asserted that is missing with federal funding. Clearly, beyond the elimination or reduction of federal funding, mitigation was no longer the number one national priority, but, rather, it was a local responsibility. This did not end hazard mitigation as a critical function, but it did weaken the promising federal–state and public–private collaborative dialogue about the national priority of sustainable community development and hazard mitigation. Soon, in conjunction with other events, it reduced a national priority to one of many things on a list of things to be done. It soon became, or so it seems, one of the less important things.

September 11, 2001, created a new and challenging dynamic for emergency management. The tragedy born of the terrorist attacks on that day understandably elevated national security concerns, but it did so in a way that many said was potentially detrimental to the emergency management function in relation to the wider array of possible natural and technological disasters. Indeed, it was even suggested that the renewed national security focus distracted emergency management and weakened its operational scope and effectiveness.

The creation of the Department of Homeland Security (DHS) established and enforced a new focus. Federal emergency management activities were reorganized under the DHS umbrella. Almost all of its traditional activities were muted by an almost exclusive emphasis on anti-terror efforts and the increased appropriations for response to the threat of terrorism. FEMA and the state and local emergency management structures across the nation were integrated into the DHS web. FEMA was placed within DHS and lost its independent cabinet-level status. The authority previously assigned to the FEMA director was vested in the DHS secretary. Not only was the stature and authority of FEMA reduced in this new arrangement, but its effectiveness and operational abilities were compromised as well (Bullock and Haddow, 2004).

With the reorganization of FEMA under the DHS umbrella, and larger allocations and an almost exclusive national focus on homeland security, concerns were immediately raised about the continuity of traditional and successful FEMA programs (Waugh, 2004). The concern that resources germane to emergency management would be diverted or prioritized for counterterrorism was great. An even greater concern was that DHS leadership and key personnel were bound to be less interested in and unfamiliar with the history, language, methodologies, and practices of emergency management. This, it was suggested, held the potential to reduce the capacity of government at all levels to deal with natural hazards as the attention of DHS was riveted on other concerns. In the initial stages, for example, it was clear that minimal attention was being paid to any matters other than terrorism-related concerns (Waugh, 2004).

Post-9-11 discussions and initiatives began to recycle many disaster myths that both scholarly research and the emergency management community had long ago rejected, including the myth of widespread panic in the aftermath of a disaster (Dynes, 2003). Likewise, the new focus on terrorism repeated the event or disaster-specific focus of the old emergency management. The DHS focus on consequence management and post-event issues placed a primary focus on first responders and command and response and too little on reducing the impact of events (i.e., mitigation) and prevention (Bullock and Haddock, 2004).

As the United States shifted its focus to preparing for and responding to terrorist-related threats and events, it was not meant of course to be at the expense of its capacity to deal with natural and technological hazards. But this had happened before. During its early days, in the context of the Cold War, 75% of FEMA's financial and human resources were dedicated to national defense. State and local governments, seduced by the prospect of FEMA funding for projects, followed suit (Bullock and Haddow, 2004). As DHS funding aimed at the threat of terrorism expanded, local communities soon emphasized anti-terror initiatives in their pursuit of federal funding. Scholars, seduced by the prospect of low hanging fruit, also chased grant money and anti-terrorism research agendas, leading to a new cottage industry of scholarly work that emphasized homeland security issues and concerns as priority one in the emergency management community. Soon emergency management was indistinguishable from national security, and the agenda seemed to be all terrorism all the time. The emphasis on hazard mitigation or other traditional avenues of emergency management scholarship was soon rivaled (perhaps even overshadowed) by the new focus on terrorism. Few scholars were discussing seriously the potential for distracting emergency management from its traditional functions by folding it into the national security agenda. The primary focus on one threat, terrorism, seemed to those who *did pay attention* to come at the expense of attention given to more frequent and widespread recurring threats.

As the DHS focus on national defense defined the federal emergency management priorities in the aftermath of September 11, FEMA became an abandoned stepchild within the DHS framework. Money for traditional emergency management activities grew tight. Additionally, both FEMA's disaster preparedness and grant-making functions were taken from it and reassigned elsewhere within the new department. Long and carefully cultivated relationships with state and local officials, not only with respect to hazard mitigation but also in regard to disaster preparedness, had seriously deteriorated, causing some to worry about the potential for poor coordination of national, state, and local efforts. By the time Michael Brown (with no previous background in emergency management) had succeeded Joseph Allbaugh (also with no emergency management background) as director, FEMA had been steadily bled to death within DHS. Many of FEMA's most experienced and able individuals had either been reassigned within DHS or had left the federal government altogether. This led to a concern about a brain drain as well as a persistent 15% to 20% employment vacancy rate within FEMA (Bergal et al., 2007). By the summer of 2005, as a result of what critics called the systematic weakening of the agency and the cut backs on traditional programs, many feared that the U.S. capacity to respond effectively to major natural disasters no longer existed. A catastrophe seemed imminent to them (Bergal et al., 2007).

Hurricane Katrina hit New Orleans on Monday, August 29, 2005. The response and recovery failures associated with Katrina are well known and have been much discussed. There is of course culpability at the national, state, and local levels. Without detailing all of the failures here, it should nevertheless be noted that the Senate Homeland Security Committee held hearings and concluded that the federal failures were directly traceable to poor leadership and a complete breakdown in the federal disaster response system (Bergal et al., 2007). In its internal review, the DHS Office of the Inspector General agreed. Interestingly, many of the ongoing problems in the recovery phase were connected to weaknesses in FEMA's grant programs that, under the new structure, were being administered elsewhere within DHS (i.e., not by FEMA) (DHS, 2006).

The White House assessment of the failures associated with Hurricane Katrina stressed an urgent need for significant changes in the status quo. It spoke of the need to adjust policy, structure, and the mindset with regard to natural disaster preparedness. This would have to involve, according to the White House report, the transformation of emergency management from every level of government to the private sector to individual citizens to local communities (United States, 2006). Without really saying it, perhaps even without knowing it for that matter, this assessment indicates that the fears of those who suspected that the integration of FEMA into DHS would result in a disaster-specific focus on terrorism to the detriment of traditional emergency management functions were justified indeed. The suspicion that the nation's capacity to respond to natural disasters had been dangerously eroded had proven to be tragically accurate. The White House assessment, calling for the transformation of emergency management, was focused primarily on disaster preparedness in its remarks. In many ways, what was being called for was the rediscovery and reconstruction of something that had once existed but had been broken. But other assessments began to raise another familiar theme. This theme, an echo of the 1990s, may be an even more critical rediscovery.

The response and recovery failures associated with Katrina, important to be sure, obscured another and perhaps more critical contributing factor to the catastrophe. Other failures less discussed may have been perhaps an even greater part of the unfolding story. The collapse of engineered systems and the systematic loss of natural defenses contributed far more to the making of a catastrophe than any of the other devastating response or recovery failures.

The entire Mississippi River Delta, including and especially the city of New Orleans, is slowly sinking into the Gulf of Mexico. This is a natural process, but it is a natural process made much worse by levees, oil and gas extraction, shipping canals, and other products of humanity's attempt to control nature. The disaster known as Katrina was not so much an act of nature as it was the act of developers, speculators, engineers, and politicians who ignored the signs of disaster resident in unsustainable development practices. It should also be noted that the engineering quality of the protections (i.e., levees) were deemed inadequate by many. But beyond that, the larger cause relates to the pursuit of development strategies that were over the long run increasing vulnerabilities. The development of New Orleans between 1965 and 2005 created conditions that were, in a word, unsustainable. Katrina was, at least in part, a human-made disaster forty years in the making.

Real estate speculators abhor a vacuum (empty land). Urban sprawl in New Orleans included the adding of levees and floodwalls to keep the water (New Orleans is surrounded by water and literally sits in a bowl beneath sea level) away from new housing developments in what had formerly been swamps. The soil in these former swamps was rich with organic materials. When it was dried out, and the organic materials decayed, the land began to sink. This means that houses, office buildings, and even flood protection levees were getting lower each year. New development on sinking land, much of it well below sea level, was protected by levees, floodwalls, and pumps. But such protection is not forever, as Karina so ably demonstrated (Freudenburg, Gramling, Laska, and Erikson, 2009). Hence, the disaster known as Katrina was not entirely an act of nature. It was also and perhaps even more so the act of the developers, the speculators, the engineers, and the politicians who ignored the signs of disaster resident in unsustainable development practices. The engineering quality of the protections (i.e., levees) was indeed proven to be inadequate, but the larger cause of the devastation that was Katrina was (for many analysts) the pursuit of development strategies that were unsustainable. The trampling of nature in pursuit of profits may be a good short-term strategy for investors, but it is the making of a disaster for the community in which it occurs. It is what happens when hazard mitigation is not incorporated into community development. It is another indication that all disasters are, alas, disasters by human design (Freudenburg et al., 2009).

To make matters worse, if that is possible, New Orleans was rapidly losing its only natural protection against hurricanes. Since 1930, New Orleans had lost 1,900 square miles of coastal land (U.S. Geological Survey, 2005). The loss of wetlands and the erosion of marshlands exposed New Orleans to the open ocean and its perilous storms as never before. This combination of a collapse of engineered systems and the systematic loss of natural defenses retaught an important lesson that was, by the end of the first decade of the twenty-first century, once again new. Among the primary lessons learned from Katrina was the need to foster community resilience. Building sustainable communities, integrating hazard mitigation into community development decisions, was rediscovered and once again recommended as a priority by many (Lansford, 2010).

Whether thinking strictly in terms of the breakdowns in response and recovery or broadening our consideration to include the failure of engineered systems and the loss of natural protections, everything that happened in Katrina was predictable. Almost everything that happened in Katrina was preventable. Most experts, including James Lee Witt, believe that FEMA and the integrated national system of the 1990s would have responded to Katrina much more effectively (Bergal et al., 2007). Beyond that, the knowledge and experience that emergency management in general had acquired over many decades, culminating in the promise of the 1990s and the imminent definition of emergency management as a profession, would have been more valuable if it had not been forgotten until it was inevitably rediscovered in a catastrophe. Emergency management might well have advanced to full professional status by now. As it is, it is picking up the pieces and attempting to reboot.

THE FUTURE OF EMERGENCY MANAGEMENT

Establishing the linkage of emergency management more emphatically to the task of building resilient and sustainable communities, it has been suggested, is an essential and necessary step for the development of an emergency management profession. This statement, articulated in our conclusion to Chapter 1 has been demonstrated in both the positive and the negative over the past two decades. The decade of mitigation, as discussed in this chapter, demonstrated both the potential for this view to shape a profession and some real progress toward its achievement. The decade adrift that followed demonstrated that the short-term horizon of decision makers, the disruptions of partisan changes in leadership, and the impact of unexpected events all too conveniently conspire to disrupt progress and interrupt intelligent discourse. A summary of the past two decades highlights the dramatic ebb and flow of the prospects for emergency management as a developing profession.

By the 1990s, largely in conjunction with national leadership from FEMA, emergency management had grown beyond a response and first responder orientation to include a comprehensive all hazards disaster preparedness, response, and recovery approach that integrated state and local efforts with federal efforts. With the Clinton administration's reinventing government initiative, and in an effort to reduce the costs of recurring natural disasters, FEMA undertook to transform emergency management by defining as its number one priority the building of sustainable and hazard-resilient communities. This step elevated hazard mitigation as the critical linkage of emergency management to the broader task of community development and mainstreamed its expertise into the nexus of local community development decision makers across the country.

As the national priority of hazard mitigation connected emergency management to the task of sustainable community development, it also provided a foundation for the redefinition of both emergency management and the foundation for a profession. The dialogue about both a new emergency management and a new emergency manager flourished in no small measure due to the connection between mitigation and sustainability. It is in direct relationship to this connection that the principles of emergency management were understood to include the anticipation of future disasters and that the defining characteristic for the emergency management professional included responsibility for taking preventive and preparatory measures to build disaster-resistant and disaster-resilient communities. The journey from trade to profession seemed to be under way, and this was also reflected in increasingly urgent discussions about the sort of educational background and training the new emergency manager would need. The transformation of emergency management, the broadening of its perspective and its role, was the result of an entirely new way of thinking about disasters. It was a fundamentally new manner of thinking. It saw disasters, or at least the damages that accompany them, less as natural events and more as the result of human decisions with respect to development, technologies, engineered systems, and the human interaction with nature. Sustainability required such thinking, and it required the special expertise of the emergency manager in the context of hazard mitigation. Sustainability required that human communities take responsibility for disasters.

With the advancement of a new national priority, and FEMA's rise as an independent agency with cabinet-level status, a national dialogue was soon under way. Through the use of policy and grant initiatives, and especially with the creation of Project Impact, this national dialogue had traction in communities across the nation. Hazard mitigation was not only incentivized by federal programs, but it was intelligently promoted by the partnerships they forged with state, local, nonprofit, and business entities. In this effort, FEMA correctly understood that changing the way we think about disasters involved doing something that has always been difficult in America's individualistic and short-term culture. It required thinking about the future, taking a long-term view that saw the necessity of acting today for benefits tomorrow. It required thinking about sustainability across the generations and not just for the span of a single generation. In promoting this new thinking at the community level, in creating a national mitigation dialogue and supporting and incentivizing local community action, FEMA was not only transforming emergency management, it was changing a culture. This may seem idealistic, and surely the national effort may be said to have had weaknesses, but it is precisely what was being undertaken.

The initial successes in hazard mitigation were promising, the dialogue was healthy, and emergency management had a purpose and a direction that seemed to be in the fast lane with respect to the development of a new profession. But priorities change, and things do happen. As the decade to follow would demonstrate, the mitigation dialogue and its potential to transform emergency management would be difficult to maintain.

With the election of George W. Bush and a new partisan orientation in the White House, it may be said that the mitigation decade ended on January 20, 2001. The new FEMA director quickly made traditional preparedness efforts a priority and supported the new administration's decision to cancel Project Impact. The national priority seemed to be simply to defund and discontinue everything the previous administration had done. Mitigation grants remained, along with the requirements of the Stafford Act, a new competitive pre-disaster mitigation grant program was in fact established, but overall mitigation funding was severely cut. Likewise, the reduced emphasis on hazard mitigation as a priority had an impact. The absence of national leadership in promoting the mitigation dialogue as national priority slowed the mitigation train considerably. September 11, 2001, brought the train to a painfully slow crawl.

With the terrorist attacks of 9-11 and the subsequent formation of the DHS, FEMA was reduced in significance and placed within DHS. Some of its functions, including preparedness and grants making, were also reassigned elsewhere within DHS. Beyond its loss of stature and the diminishing of its role, FEMA became more oriented to national security concerns. As concerns were growing that traditional emergency management functions were being underemphasized and underprioritized, communities across the country following the DHS lead placed terrorism ahead of other concerns. Mitigation was not dead, but its pulse was weak. As Hurricane Katrina demonstrated, the same can be said for traditional preparedness, response, and recovery operations as well.

Emergency management at the national level was no longer viewed as an independent or separate operation. It was a weakened and inferior version of its former self residing within a national security apparatus that had none of its experience over the past three decades since the creation of FEMA. DHS ignored much of what had been learned, ignored an existing body of knowledge, repeated past mistakes, and invited a disaster-specific focus that elevated terrorism to the detriment of other natural and technological hazard preparedness activities and, in the process, allowed previously strong federal, state, and local relationships and partnerships to deteriorate. In this context, absent a strong national entity taking the lead in traditional emergency management operations, one might suggest that the nation's attention drifted away from the critical details and the cumulative wisdom of lessons previously learned. It is also fair to say that the movement of emergency management from craft to profession also began to drift in this context.

To be sure, the discussion of emergency management as a developing profession continues. Much of that discussion was highlighted in Chapter 1. But with the folding of the national efforts into the national security umbrella and the abandonment of hazard mitigation as a national priority around which to define emergency management, the discussion is rudderless and missing a unity of theme or purpose. Some, including the Director of FEMA during the Katrina catastrophe, Michael Brown, believe neither FEMA nor the nation has improved its traditional capacities to respond to a major national disaster because FEMA remains buried within the massive DHS bureaucracy (Bergal et al., 2007). The same may be said for the prospects of transforming emergency management into a profession. But there may be other indicators to suggest that this is too pessimistic a view.

There are some indications that a more independent FEMA and a refocusing of it on broader emergency management activities, the James Lee Witt model if you will, is possible. In late 2006, Congress did in fact order a remake of FEMA within DHS, giving it greater autonomy such as that given to the Coast Guard and the Secret Service (Bergal et al., 2007). A reconsolidation of many functions within FEMA is under way. President Obama's FEMA administrator, Craig Fugate (the first since James Lee Witt to have an extensive emergency management background), has also reestablished the centrality of both hazard mitigation and the participation of whole communities in mitigation planning. Calling hazard mitigation one of the "cornerstones" of emergency management, and with an emphasis on community involvement, Fugate seems to be at least attempting to move the mitigation train and the dialogue forward (Schwab, 2011). He wants the entire village to be involved, but he also understands the need for national leadership on promoting the dialogue.

Even during the height of the drift and the distractions of the past decade, emergency management never entirely abandoned the notion that the objective of preventing disasters should permeate its actions to promote safer living conditions. It continued to regard planning for a sustainable future as part of its mission (FEMA, 2003). But the key ingredient of national leadership was inconsistent, and the reformulation or transformation of emergency management as a profession defined by its critical role in creating sustainable communities was interrupted. As a consequence, emergency management remained a collection of many occupations or trades in search of a unifying principle powerful enough to establish its professional identity and create its cohesive structure. The promise of the 1990s in this regard seemed to vanish as the new century began, causing some to lament opportunities lost. Yet the future does not exclude the possibility of regaining that promise and fulfilling it. What has been learned over time remains to guide us, what we need to do is clearly known, and the marriage of emergency management to sustainable community development can still be consummated.

A substantially new manner of thinking about where and how we want to live is, however imperfectly, becoming more evident as a need. Sustainability is an increasingly important word in the American and global vocabulary of the twenty-first century. It is necessary to promote and extend this new manner of thinking in order to maximize the potential for both sustainable development and hazard mitigation. The new manner of thinking includes the growing awareness that vulnerability is a function not of nature alone but of human behavior as well. This is evident within the emergency management community and beyond, and it must inform every step emergency management takes en route to becoming a profession.

Vulnerability in the emergency management context means the degree to which socioeconomic systems and physical assets are susceptible to natural or human-made disasters. The factors that determine vulnerability include the condition of the human populations, the infrastructures, the economies, the wealth, and the natural resources of communities. The sustainability of any community is related to the health of its three basic spheres of existence: economic, social, and environmental. Sustainable communities thrive from generation to generation because their social foundations value and are supported by a diverse inclusive economy and a vibrant ecological system that creates and sustains the opportunities for self-sufficiency and long-term security. This includes both the recognition of the limits of available resources and the avoidance of unsustainable practices that waste them for short-term gain. Sustainable communities also create and maintain a culture that avoids or mitigates natural and human-made hazards.

Sustainable communities are resilient before the stresses of extreme natural, technological, and economic disasters. Their lifeline systems of roads, utilities, bridges, and other support systems are designed to withstand the rising of water and the shaking of ground. They withstand strong winds. Their constructed environment is retrofitted to meet the threats and reduce the vulnerabilities imposed by natural hazards. Their economic and social policies and practices avoid unsustainable actions that create ruin and disaster. Into this mix, hazard mitigation (i.e., emergency management) is a critical component that contributes to the immediate goal of hazard resilience and to the broader goal of community sustainability. Strengthening buildings and infrastructure exposed to natural hazards, avoiding hazard areas for new development or redevelopment, maintaining the protective features of the natural environment, identifying and managing risks and vulnerabilities, educating the policymaker and the general public to increase their awareness of risks and mitigation strategies, and mainstreaming hazard mitigation into all community planning and development activity (public and private) are all necessary (i.e., absolutely essential) for sustainability. This necessity connects emergency management to sustainable community development, and, as such, sustainability is the principle around which emergency management must be organized.

Our national experience has demonstrated the practical benefits associated with hazard mitigation. These include reducing the loss of life and damage to property, the reduction of vulnerability to future disasters, the saving of money on recovery costs, and the benefits to community health and public safety. These are important and, naturally, should be emphasized repeatedly. But it is the broader connection to the building and maintaining of sustainable communities that these benefits contribute to which must be emphasized as the driving force for both the function of hazard mitigation and the professional development and maturation of emergency management. This is its mission and the defining principle that places the work of its many occupations and trades into a coherent whole. It makes all of the parts intelligible and their work a part of a totality that is in fact greater than the sum of its parts.

As the mitigation journey continues beyond the decade of drift and the conversation is resumed, with renewed vigor it is hoped, national leadership will be important. This does not necessarily mean a new federal program or another Project Impact, although nothing should be ruled out in that regard either. But it does mean a dialogue making mitigation once again a national priority and igniting discussion and action in communities across the country. There are healthy indications that FEMA has begun to move in this direction under the Obama administration's FEMA administrator, Craig Fugate. This is to reestablish the primacy of the goal of creating a mitigation culture. Perhaps the best way to do that is to relabel mitigation culture as a necessary component in a sustainability culture. This at once more accurately describes what mitigation and emergency management are about and invites all community decision makers and entities (public and private) to a more inclusive conversation about how, where, and why we shall choose to live. It places the goal of creating sustainable communities on the agenda as a top priority for one and all. This is the best way to create or promote a fundamentally new way of thinking so that our communities may be truly sustainable.

As progress is made in the creation of a culture of sustainability, and as hazard mitigation places emergency management at its center, the basis for establishing a professional identity for emergency management will only be aided. The practical benefits of mitigation are already known and are promoted within the emergency management community. The broader understanding of the connection between hazard mitigation and sustainability will aid in the development of an enduring and consistent emergency management mission and define the professional criteria for the achievement of it. This is where the emergency management profession will be fully born, in the consummation of the marriage between hazard mitigation and sustainability.

REFERENCES

Anderson, M. (1995). Vulnerability to Disaster and Sustainable Development: A Generational Framework for Assessing Vulnerability. In M. Munasinghe & C. Clark (Eds.), *Disaster Prevention and Sustainable Development: Economic and Policy Issues* (pp. 41–59). Washington, DC: The World Bank.

Beatley, T. (1995). Planning and Sustainability: A New (Improved?) Paradigm. *Journal of Planning Literature, 9,* 383–395.

Bergal, J., Hiles, S. S., Koughan, F., McQuaid, J., Morris, J., Reckdahl, K., & Wilkie, C. (2007). *City Adrift: New Orleans Before and After Katrina.* Baton Rouge: Louisiana State University Press.

Billiterri, T. J., & Mattingly, P. (2010). Financial Bailout. In, *Issues for Debate in American Public Policy* (pp. 245–273). Washington, DC: CQ Press.

Britton, N. R. (1999). Whither Emergency Management? *International Journal of Mass Emergencies and Disasters, 17*(2), 223–235.

Bullock, J. A. & Haddow, G. D. (2004). The Future of Emergency Management. *Journal of Emergency Management, 2*(1), 19–24.

Cigler, B. A. (1995). Coping with Floods: Lessons From the 1990s. In R. T. Sylves & W. L. Waugh (Eds.), *Disaster Management in the U.S. and Canada* (pp. 191–213). Springfield, IL: Charles C Thomas.

DHS. (2006, March 31). *A Performance Review of FEMA's Disaster Management Activities in Response to Hurricane Katrina.* Washington, DC: Department of Homeland Security, Office of Inspector General, Officer of Inspections and Special Reviews.

Dynes, R. (2003). Finding Out in Disorder: Continuities in the 9-11 Response. *International Journal of Mass Emergencies and Disasters, 21*(3), 9–23.

FEMA. (1997). *Report on Costs and Benefits of Hazard Mitigation.* Washington, DC: Federal Emergency Management Agency Mitigation Directorate.

FEMA. (1999). Project Impact: Building a Hazard Resistant Community. Washington, DC: FEMA release 1293-71 at http://www.fema.gov/news/newsrelease.fema?id=8895. Accessed June 5, 2012.

FEMA. (2003). *Planning for a Sustainable Future: The Link Between Hazard Mitigation and Livability.* Washington, DC: FEMA, Publication 364.

Freudenburg, W. R., Gramling, R., Laska, S., & Erickson, K. T. (2009). *Catastrophe in the Making.* Washington, DC; Covelo, London: Island Press/Shearwater Books.

Galvin, K. (2001, March 2). Embarrassing day for Project Impact cut. *The Seattle Times.*

Godschalk, D. R., Berke, T., Brower, D. S., & Kaiser, E. J. (1999). *National Hazard Mitigation: Recasting Disaster Policy and Planning.* Washington, DC: Island Press.

Holdeman, E. (2005). Destroying FEMA. *Washington Post.* http://www.yuricareport.com/Disaster/DestroyingFEMA.html. Accessed June 5, 2012.

Holdeman, E., & Patton, A. (2008). Project Impact Initiative to Create Disaster Resistant Communities Demonstrates Its Worth in Kansas Years Later. http://www.emergencymgmt.com/disaster/Project-Impact-Initiative-to.html. Accessed June 5, 2012.

Kates, R. (2003). Making the Transition. *Government Technology, 16*(4), 10–15.

Kolva, R. (2002). Effects of the Great Midwest Flood of 1993 on Wetlands. U.S. Geological Survey. http://water.usgs.gov/nwsum/WSP2425/flood.html. Accessed May 18, 2012.

Kusterer, K., Ruck, M. T., & Weaver, J. H. (1997). *Achieving Broad-Based Sustainable Development: Governance, Environment, and Growth with Equity.* Hartford, CT: Kumarian Press.

Lansford, T. (2010). *Fostering Community Resilience: Homeland Security and Hurricane Katrina.* Farnham, Surrey, England: Ashgate.

Mooney, C. (2007). *Storm World.* New York: Harcourt Brace.

Mileti, D. (1999). *Disasters by Design: A Reassessment of Natural Disasters in the United States.* Washington, DC: Joseph Henry Press.

Osborne, D., & Gaebler, T. (1993). *Reinventing Government: How the Entrepreneurial Spirit Is Transforming the Public Sector.* New York: Addison-Wesley Publishing Company.

Schneider, R. O. (2002). Hazard Mitigation and Sustainable Community Development. *Disaster Prevention and Management, 2*(1), 25–29.

Schwab, A. J., Eschelbach, K., & Brower, D. J. (2007). *Hazard Mitigation and Preparedness: Building Resilient Communities.* New York: Wiley.

Schwab, J. C. (2011). Hazard Mitigation: Integrating Best Practices into Planning. American Planning Association. http://wyohomelandsecurity.state.wy.us/pubs/haz_mit_png_best_practices.pdf. Accessed May 29, 2012.

Sylves, R. T. (1996). Redesigning and Administering Federal Emergency Management. In R. T. Sylves & W. L. Waugh (Eds.), *Disaster Management in the U.S. and Canada* (pp. 5–25). Springfield, IL: Charles C Thomas.

United States. (2006). *The Federal Response to Hurricane Katrina: Lessons Learned.* Washington, DC: White House.

U.S. Geological Survey. (2005). Depicting Coastal Louisiana and Loss. Fact Sheet 2005-3101. http://www.nwrc.usgs.gov/factshts/2005-3101.pdf. Accessed May 22, 2012.

Waugh, W. L. (1990). Emergency Management and the Capacities of State and Local Governments. In R. T. Sylves & W. L. Waugh (Eds.), *Cities and Disasters: North American Studies in Emergency Management* (pp. 221–237). Springfield, IL: Charles C Thomas.

Waugh, W. L. (2004). The All Hazards Approach Must Be Continued. *Journal of Emergency Management, 2*(1), 11–12.

Woolsey, M. (2008). America's Most Expensive Natural Disasters. *Forbes.* http://www.forbes.com/2007/10/29/property-disaster-hurricane-forbeslife-cx_mw_1029disaster.html. Accessed May 18, 2012.

Chapter 3

EMERGENCY MANAGEMENT ETHICS AND SUSTAINABILITY

Human History is a race between education and catastrophe.

– H. G. Wells

Integrity without knowledge is weak and useless, and knowledge without integrity is dangerous and dreadful.

– Samuel Johnson

INTRODUCTION

Health and safety have, in recent times, joined goodness, truth, and justice among the pantheon of Western culture's defining principles. Health and safety can also be said to represent imperatives for emergency management generally (Partridge, 1988; Schneider, 1993; Shue, 1981). But competing and legitimate interests, together with the internal contradictions they present, may generate destructive outcomes as easily as fruitful ones. Whether in relation to emergency management, sustainable development, or economic, social, and political life generally, decision making is challenged to identify and adhere to guiding principles that promote fruitful outcomes while discouraging destructive ones. Absent a set of guiding principles that protect against it, our clever and advanced species can quickly and efficiently justify and implement actions that place health and safety at risk as we pursue what we have convinced ourselves are necessary advances made possible by the great and mighty deeds our vast knowledge and ever evolving technology make possible. But as knowledge and technology increase our capacity to accomplish great and mighty deeds, there is a corresponding need (duty some might say) to expand our notion of responsibility for their outcomes and broaden our ethical thinking.

Sustainable development and hazard mitigation, as a vital component in sustainable development, are concepts that expand the notion of our responsibility for outcomes. They are by their very definition a broadening of our ethical thinking. As these concepts were discussed throughout Chapter 2, an inevitable ethical argument was already being formed. It was noted there that sustainability is the effective use of resources (natural, human, and technological) to meet today's needs while ensuring that these resources are available to meet future needs. Sustainable development thus is the meeting of today's needs by communities without compromising the ability of future generations to meet their needs. Hazard or disaster mitigation, in this context, is the fostering of sustainable communities in the face of the risks and vulnerabilities that potential and extreme hazardous events inevitably impose on them. This, as we shall argue in this chapter, presents us with the guiding principle that defines both the fruitful outcomes we wish to promote and the destructive ones we wish to avoid through our marriage of emergency management in general to the broader concept of sustainability.

If emergency management is to blossom into a profession, it must, as all professions do, identify the ethical standards and codes that represent its core values and guide its practitioners in the many individual and specialized aspects of their work. As the ongoing discussion of emergency management as a profession has progressed, so too has the discussion of ethical principles for a developing profession. But both of these discussions remain incomplete. A paucity of scholarly work on ethics in emergency management makes it difficult to adequately consider the ethical foundations of emergency management policy. It also makes the development of the ethical principles for a profession less than definitive. Should the movement toward defining the profession in the context of sustainability continue and establish a foundation for it, the threads of ethical theory already available and in use can evolve into a more inclusive statement of ethical principles for emergency management. As we begin to travel toward that more inclusive statement of principles, let us first discuss what has already been developed and applied to the existing ethical standards for emergency management.

EMERGENCY MANAGEMENT ETHICS
AND CURRENT STANDARDS

In general, for at least the past twenty-five years, it has been established and agreed on that emergency managers are ethically responsible under a specific set of conditions. These conditions are broadly articulated as follows:

- they have knowledge of or are able to predict an emergency or disaster;
- they have the capacity of making a decision and acting upon it;
- they have a choice, i.e., they could have chosen otherwise; and
- they make decisions that have value consequences affecting the safety, lives, welfare, and rights of other persons (Partridge, 1988).

An analysis of these "conditions" has been the traditional place from which to begin a discussion about the nature and scope of ethical responsibility for emergency managers. These conditions are frequently discussed and interpreted, within the United States, in the context of the basic moral criteria associated with the traditions of Western thought.

Among the basic alternatives in the Western tradition for moral criteria in relation to emergency management ethics are: utilitarian rationales, the concept of basic human rights, culpability and prevention of harm standards, the imperative of knowledge, and public service rationales (Beatley, 1985; Lilla, 1981; Schneider, 1993). Let us briefly examine each of these alternatives and highlight their implications as moral criteria for ethics in emergency management. Without any suggestion of choosing from among these alternatives, it is easy to see that each of them has had some impact on ethical thinking in relation to emergency management as well as some influence on existing ethical codes that have been developed.

Utilitarian philosophers such as Jeremy Bentham and John Stuart Mill evaluate the ethical desirability of an action based on its usefulness for creating the greatest good for the greatest number (Bagby, 2002). This greatest good is typically defined in terms of the material and non-material things that pleasure the individual. With respect to public policy in relation to hazards and disasters and related emergency management functions, the preferred or ethical action is that which maximizes net social benefits (i.e., defined as the collective accumulation or aggregation of individual benefits) at the lowest social cost (Beatley, 1985). The utilitarian approach has been institutionalized in the public sector through the implementation of cost-benefit analysis. Whether practicing emergency managers are aware of it, utilitarian cost benefit-analysis drives many of the policy decisions that shape the contours of their work. Of course utilitarianism has its critics and its limitations as a foundation for decision making. At what point, for example, does the social cost (think taxes perhaps) become great enough (i.e., exceeding the benefits however defined and quantified) to justify the toleration of increased risks, including life-threatening ones, that place the general public at greater danger? The utilitarian calculus, to the extent that its focus might be narrowed to immediate costs and benefits, may easily justify some outcomes that would be dangerous or unacceptable in the long run. Thus it is that the utilitarian calculus is often tempered by some notion of basic human rights, including the right to safety.

The basic human rights formulation suggests that every individual has certain basic or inalienable rights, including the right to physical security (Shue, 1980). This means individuals may be said to have the right to a basic minimum level of public and personal safety and that it is never morally justifiable to allow a significant loss of life from any predictable disaster without taking action to prevent or minimize it. Even the argument that it would be socially inefficient (i.e., too costly) does not justify refusing to act in a manner required to maintain public safety (Beatley, 1985).

The basic human rights argument is compatible with, and an extension of actually, the Lockean (i.e., John Locke) concept of life as a property right and the associated notion that neither governmental nor private entities may violate or allow to be violated by others the "lives, liberties, and estates" of citizens (Locke, 1965). It also embodies perfectly the Jeffersonian notion of life as one of the inalienable rights that serve as the foundation for American culture. Given the value placed on human life, the saving of lives and the prevention of harms or human suffering would be chief among the moral objectives of emergency management (Schneider, 2002). This argument often includes the recognition of the fact that the impact of devastating natural disasters, for example, is greatest among the poor and disadvantaged populations. This fact suggests the moral and practical necessity of public policy and public intervention to save lives.

The basic right to protection from disasters leaves undecided the status of the protection of property. That is a separate matter. The basic questions regarding the protection of property, the protection against economic displacement, and the preservation of lifestyle are sustainability issues that are factored into risk calculations, policy decisions, preparedness and mitigation planning, but they are not part of the basic human rights argument (Beatley, 1985; Schneider, 2000). They are, as we have seen in Chapter 2, critical aspects of sustainable development in every community. To the degree that sustainability is articulated as a social value and a priority, they do take on both a moral and an ethical significance in the context of the social contract. But the basic human rights argument stems from the Western valuation of life as a natural right. This valuation of life leads, in fact, to the development of prevention of harm or prohibited risk standards.

Beginning with the premise that the basic right to life includes the variable of personal safety, it follows that the preservation of life and the prevention of avoidable harms must figure into every moral calculation of risk. Risks are defined as the threat of harm or loss in relation to things of value (e.g., life, health, vitality, etc. in this case). Risks that may be considered to be prohibited by the basic human rights argument are defined along the following lines: the potential harm is physical and life threatening, the potential harm is undetectable by potential victims, the potential harm is predictable, the potential harm is preventable, and the probability of incurring the harm is high (Schneider, 1993; Shue, 1981). Under these sets of circumstances, the risk is prohibited (i.e., it may not be taken or tolerated by policymakers and public actors). The concept of prohibited risk, in this formulation, becomes a moral imperative for emergency policymakers and managers because individual citizens or potentially impacted populations cannot usually perceive or predict a threat to life or safety and pursue their own best interests (however defined) in a complex hazard scenario.

At its heart, the notion that it is not allowable for one party or parties (public or private) to inflict the risk of damages and loss of life onto other individuals or onto the public at large is commonly accepted in both the Lockean right to life formulation and in John Stuart Mill's classic treatise *On Liberty*. Much of what has evolved in emergency management policy and implementation activity, especially with regard to preparedness, response, and mitigation, is in fact compatible with this rationale. One can see its application in the guise of preventing harms (mitigation) in the form of protecting people, communities, economies, and constructed structures against the devastation associated with a natural or human-made disaster that may pose a threat to life and the conditions necessary to support and sustain it.

It is worth taking special note that the prevention of harm, the prevention of prohibited risk, even a basic utilitarian cost-benefit analysis all have one important thing in common, and this is a critical component in any formulation of the ethical responsibilities for emergency managers. The meeting of any professional responsibilities, ethical or other, is based on the assumption of knowledge and a central role for it. This is no small ingredient in the mix.

Emergency managers and policymakers alike must know present situations, predict risks and harms, develop appropriate technical and organizational responses, anticipate outcomes, and be forward looking with respect to reducing threats to human life and safety. The development of knowledge, especially predictive or anticipatory knowledge, is a requirement that should be perceived by emergency managers as a professional duty. Why? Without the development of such knowledge, regardless of the moral criteria employed, there can be no basis for ethical responsibility (Schneider, 1993, 2000). In fact, it could be said (much as it can be said of the medical profession, for example) that knowledge is a moral imperative for ethical responsibility in the field of emergency management. Whether maximizing social benefits in some purely utilitarian calculus, identifying and preventing harm in a basic human rights calculus, or contributing to sustainable community development generally, knowledge is a requirement or criteria for ethical (i.e., responsible) action. As we shall see, most current formulations of ethical standards incorporate this understanding about the importance of knowledge as a professional requirement.

It should finally be noted that the literature of ethics in public administration has also been applied to the emergency manager as a public actor. Ethical analysis in the contexts of public service (Garofalo, 1999; Rohr, 1998), public administration (Denhardt, 2003), and public integrity (Dobel, 1999) has utility when applied to efforts to putatively define the emergency manager as a professional and emergency management as a profession. The cultivation of responsibility for public resources and public well-being, serving the public interest, the improvement of the moral and cognitive capacities of public managers, the creation of ethical awareness, and the development of responsibility are all components of a public service ethic that applies to the emergency manager as a public servant.

Most of what we have discussed up to this point has, in one form or another, quite naturally influenced the ongoing attempts to define the ethical responsibilities of emergency managers and emergency management as a putative profession. Progress has been made to be sure. Most state and local emergency management associations, following the lead of the IAEM code of ethics, have fashioned fairly similar ethical standards and codes that reflect the basic principles we have been discussing. They follow a format that utilizes an agreed-on formula of basic values and leads to the identification of basic, albeit general, ethical principles. Insofar as these codes go, they do embody most of the traditional Western foundation we have laid out in our discussion. But they also seem to be lacking in the sense that their generality has not really contributed to the more precise definition of emergency management as a profession. Before assessing what may be lacking, let us examine the existing codes.

The codes of ethics adopted by IAEM and many state emergency management associations adhere to a set of core values: respect, commitment, and professionalism. These core values are presented in ethical codes meant to reflect the spirit and conduct of emergency management activities directed by the conscience of society (i.e., the social contract) and dedicated to the well-being of all. They are said to constitute the standards for ethical and professional conduct generally. What follows is a representative sample of these ethical codes as it appears in the Alabama Emergency Management Association's (AAEM) code of ethics. It is based on the core values of respect, commitment, and professionalism.

- Respect – Respect for supervising officials, colleagues, associates, and, most importantly, the people we serve is the standard for AAEM members. We comply with all laws and regulations applicable to our purpose and position, and we responsibly and impartially apply them to all concerned. We respect fiscal resources by evaluating organizational decisions to provide the best service or product at a minimal cost without sacrificing quality.

- Commitment – AAEM members must commit themselves to promoting decisions that engender trust from those we serve. We commit to continuous improvement by fairly administering the affairs of our positions, fostering honest and trustworthy relationships, and striving for impeccable accuracy and clarity in what we say or write. We commit to enhancing stewardship of resources and the caliber of service we deliver while striving to improve the quality of life in the community we serve.

- Professionalism – AAEM actively promotes professionalism to ensure public confidence in emergency management. Our reputation is built on the faithful discharge of our duties. Our professionalism is founded on education, safety, and the protection of life and property.

Each state utilizes pretty much the same language in articulating the principles of respect, commitment, and professionalism into their codes. Let us briefly examine this language in relationship to the various criteria we have been discussing with respect to ethical standards in emergency management.

The value of respect includes much of the language associated with the public service criteria (especially public integrity) and emphasizes the conduct requirements for all public actors and agencies that interact with other public individuals and organizations. Likewise, the emphasis on responsible management of public resources and accountability to the public is emphasized. The language about providing the highest quality of service at a minimal cost is ripe for or at least suggests the application of the utilitarian criteria for the maximization of social benefits.

The value of commitment emphasizes other public service concerns (trust, honesty, stewardship, etc.) but can also be said to introduce the necessity of a knowledge-based criteria (i.e., impeccable accuracy and clarity). Nevertheless, the primary focus here is still on the administrative and public service component of the work.

The value of professionalism, with its added emphasis on safety, protection, and the preservation of life and property, connects most directly to the basic human rights criteria and the inferred rights of personal safety associated with it. Once again, and this time more directly, education is emphasized and knowledge is alluded to as a critical criterion.

The connections noted between the existing ethical codes and various ethical criteria we have discussed are not, of course, explicitly detailed. The agreed-on values of respect, commitment, and professionalism incorporate them perhaps as a result of cultural habit (i.e., these notions are embedded in Western and American political thought). These codes reflect basic themes in American culture and in the public service literature, but they are almost too generic to fuel the more precise defining of an emergency management profession. While good general statements, one cannot help but feel that these codes need to be more explicitly tied to the mission of a profession and to guide more precisely the promotion of professional and ethical conduct. We now turn our attention to a discussion of this need.

ETHICS AND SUSTAINABILITY: THE CRITICAL LINK

A suggested and desirable axiom is that ethical responsibility is a correlate of power, and the exercise of power demands foresight (Jonas, 1984). It is most wise to remember this axiom in relation to the performance of all public functions, including emergency management. The notion of foresight, a future orientation if you will, plays a critical role in the connection of emergency management ethics to sustainability.

At its heart, ethical responsibility in a public setting must be more than the proscribing of behavior with respect to immediate relationships and activities. It must be, as a correlate of power, future-oriented. It demands foresight and responsibility for long-term outcomes. As most policy and public management experts can tell you, this foresightedness or future orientation is hard to come by. It often requires a predictive knowledge that is not always cultivated or available, and it is frequently frustrated by the immediacy of events and short-term objectives that may skew perceptions. But emergency managers might, should they contemplate the matter, be among those most likely to comprehend the need for a future orientation in their work.

In considering the ethical dimensions of emergency management, one is apt to be frequently caught between the known and the unknown. Public service in this field performs a function (i.e., exercises a power) that requires decision making and responsibility in relation to outcomes and impacts that are problematic. They must identify risks and vulnerabilities related to natural, technological, and a variety of other hazard potentials that threaten their communities. They must predictively anticipate hazardous events, be prepared to respond to them, and recommend and design measures to mitigate them. In the meeting of their work-related responsibilities, as it shall be argued herein, knowledge plays a special and central role in their work. It is a necessity in the practical application of the various crafts that comprise its efforts that emergency management bridge the gap between the known (quantifiable risks and vulnerabilities, predictable natural or human-made hazards) and the unknowable (what will happen, when it will happen, how severe it will be, and what its impact will be with respect to the sustainability of the community). Bridging the gap is also a prerequisite for the meeting of any ethical responsibilities one might define for emergency management. The same may be said with regard to the concept of sustainability. There is no sustainability without bridging the gap between the known and the unknown. Perhaps most interestingly, *sustainability* may be said to be precisely what the exercising of power in any field is about to the degree that it is indeed an ethical undertaking (i.e., it is the principle that promotes fruitful (sustainable) outcomes while discouraging destructive (unsustainable) ones).

Sustainability (i.e., the effective use of natural, human, and technological resources to meet today's needs while ensuring that these resources are available to meet future needs), is by definition future-oriented. It requires the anticipation and management of threats and vulnerabilities that deprive future generations of the opportunity to meet their needs, and it requires the avoidance or elimination of unsustainable practices that fuel those threats and vulnerabilities. This requires, among other things, the cultivation and application of an informed knowledge-based foundation for decision making. As a value, sustainability may be said to be a foundation for ethical principles that define both desired outcomes and the criteria for responsible action. In effect, unsustainable practices that knowingly and willfully threaten sustainability would be considered unethical by the presupposed social value attached to or implicit in the concept of sustainability. Unsustainable practices would constitute the frustration or avoidance of the stated responsibility to ensure that future generations will have the ability to meet their needs.

As a core value, sustainability represents a moral foundation for social, political, economic, and human survival. Whether considered in the context of a basic and practical utilitarian rationale or in the broader context of basic human rights, it can also be said that sustainability as a value is compatible with the traditions of Western thought. To the extent that emergency management may become defined as a profession as a sustainability enterprise, its core value and ethical dimensions would be clarified. This applies to all phases of emergency management. While Chapter 2 focused on hazard mitigation as a necessary and critical component in the building of sustainable communities, the emergency management phase that brings emergency management emphatically into the broader orbit of community development, it may also be said that disaster preparedness, disaster response, and recovery activities are also connected to the sustainability of communities. If this is so, then sustainability as a value is the ultimate ordering principle for all activities within the emergency management realm.

Sustainability, as a value, is a container that holds all of the various traditions of Western thought quite nicely. The protection of life and the prevention of harm associated with the basic human rights perspective require or at least imply a forward-looking approach that sustains both life and the conditions necessary to support it. Likewise, the utilitarian rationale if applied with respect to long-term costs and benefits makes sustainability a necessary part of its calculations to the extent that it may recognize that the future impacts of today's decisions and actions hold the potential for knowable and predictable long-term costs and benefits. Professional responsibilities associated with the responsible use of public resources, the relationships with organizations and the general public, and the development and applications of knowledge and public safety also imply a future orientation and emphasize the implementation of sustainable practices. In a sense, one can suggest that sustainability is a value at the core of *all* moral traditions in that they seek to promote over time the practices and behaviors that promote and preserve desired outcomes and to discourage those that produce undesirable ones.

With respect to both sustainability and emergency management, and quite independent of the acceptance of the contention that emergency management is (putatively at least) a sustainability-oriented profession, the development and application of knowledge is the place where ethical considerations would seem to necessarily begin. The nature and scope of sustainability as a goal and of emergency management as a practice are such that the ethical context for both moves beyond direct and immediate dealings with people (this is where much ethical thinking often begins and ends) and involves decisions and actions that have a causal reach into the future. This being the case, the fact that we are discussing intergenerational relations and impacts, the development and application of knowledge, the ability to predict, the special nature of public responsibility, and the long-term consequences of action and/or inaction must figure strongly into any examination of ethical responsibility. So it is with the role of knowledge that the development of emergency management ethics, whatever its organizing principle, must begin.

The emergency management literature has spoken from its beginning of the need to anticipate the unexpected and reduce the risks to life, safety, and property posed by regularly occurring natural hazards or by hazardous events stemming from human endeavors that entail elements of predictable risks (Petak, 1985). This implies both a forward-thinking or future orientation and a commitment to the reduction of risk as a primary function of emergency management. As we have already noted, the early literature emphasized that emergency managers and public decision makers are ethically responsible for the outcomes of their action or inaction when they have knowledge of hazard risks and are able to prevent disasters, they have the ability to make decisions, they have a choice to act or not to act, and their decisions have value consequences (Partridge, 1988). What is to be emphasized now is the special consideration of knowledge and its essential connection to ethical action. A proper understanding of knowledge and its role leads, inevitably some might suggest, right back to the concept of sustainability as a defining characteristic of the work in the field of emergency management.

Policymakers and emergency management specialists presumably operate on the basis of knowledge about present situations, projected risks, and possible harms as they develop responses and strategies that anticipate future outcomes. This requires a knowledge base that is predictive and future-oriented. It is perhaps an unfortunate or inconvenient truth that our ability to act, in a technological sense at least, may sometimes exceed our predictive capacities. This gap between our ability to act and our ability to predict, to the extent that it may in fact exist, is a moral dilemma in that it is ultimately impossible to define responsibility for the unknown. Narrowing the gap between our ability to act and our ability to predict, between the known and the unknown, is a necessary precondition for ethical responsibility. Without the proper knowledge base or the ability and willingness to use it, there is no possibility for ethical choice no matter what the value or values being served by that choice. Whatever its causes, the lack of a proper knowledge base or a refusal to incorporate it into applied action will render all decisions arbitrary from an ethical point of view. Ethical judgments pertaining to both emergency management and sustainability, improperly informed or based solely on the pragmatic discretion of decision makers, are no better than whim or impulse when they are not founded on reliable and predictive knowledge. It is important to note, however, that the lack of such knowledge is often a result of unethical choices already made.

The unavailability of knowledge due to a genuine inability to develop reliable predictive information is to be truly confronted with the unknown, and it renders predictability impossible. This inevitably suggests a tragic ethical arbitrariness and a degree of uncertainty that precludes holding any relevant actor or decision maker responsible for unexpected outcomes. However this condition, the inability to know, is infrequent and rare. More typical is the unavailability of knowledge due to ignorance, a condition that is correctable. In this instance, the possibility of knowing exists but has not been fully developed or pursued. Not only is the problem correctable, it may also be considered a duty or an obligation to take every step necessary to close the gap between our ability to act and our predictive awareness of consequences. There is a third type of unavailability with respect to knowledge, and this unavailability represents the true definition of unethical behavior with respect to the work of emergency managers and in relation to the value of sustainability as well.

The third type of unavailability with respect to knowledge is at once the most prevalent and the most damaging type with respect to both sound emergency management and sustainability. This is the unavailability of knowledge by choice. Unavailability by choice refers to a conscious decision to ignore or disregard what is known in order to pursue objectives that, while often profitable and beneficial in the short term, reliable predictive knowledge reveals to be damaging and unsustainable in the long run. Examples of this phenomenon abound. The decision of an industry to cut safety-related costs in order to maximize profits may knowingly place workers and often the general public at greater risk. The intentional ignoring of threats to the environment or the ecosystem that results in increasing the risks to human populations and reducing the long-term sustainability prospects for communities in the pursuit of more immediate self-interested objectives (i.e., cutting costs, increasing profits, etc.) is another common pattern. The promotion of high-risk development that is profitable in the short term but destructive of sustainability in the long term is another. The decision not to retrofit buildings or spend resources on other mitigation techniques that will save money in disaster recovery scenarios simply to avoid the immediate expenditure of money is another common example. On and on we could go. In all such cases, there is a conscious decision to either roll the dice on long-term risk or knowingly load up on too much unsustainable long-term risk for immediate short-term gain.

The unavailability of knowledge by choice is almost always the product of political or economic influences that induce decision makers to discount, underestimate, ignore, or deny what is known about risk in pursuit of what they perceive to be more immediate and compelling political and economic values. Their focus is on neither the future and sustainability nor the reduction of risks and vulnerabilities. This does not mean that a focus on immediate and legitimate political or economic objectives is an entirely bad or unnecessary thing. It certainly does not mean that decision makers wish to cause great harm to either present or future populations in a community. It simply means that they choose to ignore some of the more inconvenient truths about the long-term impact of their actions or to claim any responsibility for them. That is ultimately the temptation most easily surrendered to with short-term thinking. There is often an irresistible allure to the immediate incentives that make unsustainable practices look less dangerous when compared with short-term benefits or profit potential. This human tendency, often encouraged by the inevitable impulses of an economic and political culture that nurtures it, is precisely why those whose work is to promote sustainability must adhere to an ethical framework that is genuinely forward thinking and future-oriented. Absent professions (both public and private sector professions) that are defined by, adhere to, and promote such a framework, there is no moderating the ill effects of short-term thinking. Promoting long-term, future-oriented thinking and action is the defining mission of any sustainability profession. Should emergency management come to articulate itself as a sustainability profession, it is the guiding principle that must inform its unfolding definition of itself and the development of its professional ethic.

With respect to knowledge, it is important to see it as a precondition, but not a guarantee, of ethical decision making. Ultimately, the values that the applications of knowledge are meant to serve and an adherence to them define the criteria for ethical choice. In articulating sustainability as the primary value defining the work of emergency management, in all of its phases, the knowledge and the techniques associated with each of its specializations has one purpose, its mission may be defined, and a profession may have a firm foundation. As discussed in Chapter 2, the hazard mitigation function ties emergency management most directly and most critically into the broader web of planning and implementing the development and maintenance of sustainable hazard-resilient communities. But the same may be said to apply generally to the anticipation of and preparedness for natural and human-made hazards and the disaster threats they present. The same may be said with regard to the efficient response to disaster occurrences and with respect to the recovery from them. *All are among the things sustainable communities must be committed to doing effectively* in meeting their responsibilities to both present and future generations. Emergency management, to whatever extent it is or it may become a profession, is best defined in connection to the value and goals associated with sustainability. The linkage between emergency management and sustainability is an unbreakable one in practice, and to the degree that sustainability may be thought of as the public value by which its work is to be evaluated, the ethical responsibilities of emergency management are connected first and foremost to the goals associated with the development and maintenance of sustainable communities.

Ideally, a more complete and comprehensive statement of the ethical principles of emergency management would include many, if not most, of the components from our present discussion. It would include the public service ethic and the building of relationships based on integrity. They would include the responsible use and management of public resources. It would include acknowledgment of and responsibility for the public and/or individual right to safety as a human right. This might include a special responsibility for those who are disproportionately impacted by natural and human-caused disasters – the poor. Among the other components included would be the reduction of risks and the avoidance of prohibited risks. Additionally, the ethical principles developed would include language related to all disaster phases, preparedness, response, recovery, and mitigation. Most importantly, the presentation of each component would be connected to the relevant moral criteria associated with the value of sustainability. Finally, all of this would be entirely consistent with the core values of respect, commitment, and professionalism already embodied in existing codes of ethics. But all of these components would take on greater meaning and make a more powerful statement if the linkage between emergency management and sustainability were understood to predominate all else. Sustainability (i.e., the development and maintenance of sustainable communities) as the purpose, the mission, and the value to be served by emergency management may provide the foundation it needs to define itself as a profession and to state more precisely its ethical responsibilities.

Given the goal of sustainability and the anticipatory nature of emergency management, the ethical principles for emergency management would emphasize forward thinking. A future orientation implies that, with respect to both the technical and human components of its various functions, any statement of ethical principles for emergency management must emphasize knowledge as an ethical imperative. As we have said, the nature and scope of emergency management and sustainability creates an ethical context that moves well beyond direct and immediate dealings with people and organizations. It involves decisions and actions that have an impact and a causal reach far into the future. Hence, knowledge of a particular kind, the ability to predict or anticipate, and an understanding of long-term consequences of immediate actions must be included in any articulation of emergency management ethics. This applies to all disaster phases but most obviously and critically to the hazard mitigation function.

The centrality of hazard mitigation for sustainable community development, with respect to the reduction of risk and the prevention of harm generally, requires the application of predictive and anticipatory knowledge. Given the economic and human costs associated with hazardous events, mitigation is a core necessity in practical (i.e., utilitarian), human (i.e., prevention of harm, right to safety, etc.), and sustainability contexts. Indeed, one might suggest that just as mitigation is the disaster phase most directly connected to sustainable development (but not to the exclusion of the other phases), so too it is most directly connected to the ethical core (again not to the exclusion of the other phases) of emergency management.

Based on our discussion of current codes and the principles of respect, commitment, and professionalism, the general discussion of Western moral criteria, and the articulation of sustainability as the proposed core value to be served, one can begin to imagine what a more comprehensive statement for ethical principles for an emergency management profession might look like. The following suggests itself as a worthy illustration (key points presented in bold).

- **Principles of Emergency Management Ethics**. Emergency managers assume specific ethical obligations that arise out of the special features of professional emergency management practice. These principles express the fundamental ethical responsibilities of emergency managers as professionals and as public servants whose leadership and work are essential for the development and maintenance of hazard-resilient communities.
- **Emergency managers shall**:
    1. **embrace the public welfare** as their primary responsibility;
    2. strive in all professional activities to **protect the best interests of all** in their communities, but particularly those most vulnerable and unable to cope with the impact of a hazard or disaster (e.g., the poor, children, the elderly, the disabled, pets);
    3. deal **fairly and honestly** with colleagues, other organizations (governmental and nongovernmental), and the public **while promoting professional competence**, informed policy, and sound practices;
    4. **act as responsible stewards** of the public resources entrusted to them;
    5. **foster hazard mitigation efforts that contribute to sustainability**, including those linked to the natural resource environment that will maintain or enhance its long-term protective features;
    6. promote the development of **hazard-resilient and sustainable communities**;
    7. **respond** promptly, expertly, and without prejudice or partiality to all community needs associated with a disaster incident;
    8. **work cooperatively** with other community leaders to ensure that emergency planning and preparedness is effective and that community development planning does not shift risks to other communities, to at-risk or vulnerable populations within the community, or to future generations;
    9. support and provide leadership as appropriate for all efforts to **build a consensus** among all groups and people having a stake in the outcomes of all hazard mitigation, planning, pre-

paredness, response, and recovery operations;
10. engage in **continuing study** and education **to maintain and/or enhance the knowledge and skills** necessary to provide high-quality emergency management services.

These ten principles would seem to be practical enough to guide the practitioner but also inclusive of some of the broader foundation needed to define the profession.

Many of the principles listed above contain elements of the public service criteria for ethical responsibility. The references to public welfare, fair and honest dealings with colleagues, organizations, and the public, the responsible use of public resources, and working cooperatively with other leaders are all the basic staples of any public service ethic. The requirement that emergency managers act as responsible stewards of public resources also may be said to invite the application of a utilitarian calculus. The prevention of harm, prohibited harm, basic rights criteria are especially apparent in several of the principles, and the critical role of knowledge in relation to the performance of duties is present as well. But it is perhaps the inclusion of the future orientation (not shifting risks to future generations) and the emphasis on hazard mitigation and sustainability that are the most compelling attributes of these principles.

The emphasis on hazard mitigation and sustainability (principles five and six) and the concept of responsibilities spanning generations and communities (principle eight) is where the future or long-term orientation of the profession is expressed. This is, on a practical level, what distinguishes emergency management as a function that unites all of its specializations and activities and gives them coherence and meaning. This may sound too idealistic to the person working in a response agency or a local emergency manager simply trying to be prepared for the next regularly and predictably occurring natural disaster. But remember, it is the profession we are attempting to define in relationship to sustainability, and this refers to the combined effect of emergency management and not simply the immediate tasks of the individual in the field. But that individual, even though he or she may be focused on a narrow slice of a particular function, does need to see and feel connected to the broader purpose. This will require more than a statement of ethical principles. It will require a definition and an unambiguous understanding of the purpose to be served. Before concluding this chapter with a broader discussion of that purpose, a final comment about ethical principles is in order.

All professions and social organizations ascribe to a set of beliefs and values. Doctors and lawyers sign an oath and commit to upholding the ethical requirements of their profession. All professions have a unified commitment to develop the most professional organizations possible. A part of that development, and especially in professions impacting lives, health, safety, and the public welfare, requires a well-constructed and universally implemented set of ethical principles that establish the standards for performance and define the responsibilities to be met. Emergency management is no different in this regard.

The existing codes of ethics are a positive sign that emergency management is striving to become a profession. But the general nature of these codes does suggest the need for a clearer connection of the work of the putative profession and the training and development of its practitioners to the service of a purpose that creates of its parts (i.e., specialized functions and trades) a whole that speaks authoritatively and broadly to promote an agreed-on public value that requires its (i.e., the profession's) leadership and the contributions of its constituent trades and practitioners. In this sense, the linkage of emergency management to sustainable development and disaster resilience becomes more than a practical matter. This relationship, more than perhaps any statement of ethical principles, spells out in great clarity the foundations of a profession and defines the scope of its ethical responsibilities. The need for such a foundation does not make emergency management unique, for all professions require one.

THE FUTURE OF EMERGENCY MANAGEMENT

Most professions have different specialists, and none of them alone is sufficient to accomplish the goals or implement the values of the profession as a whole. The medical profession, for example, has many specializations and technicians. Individually they do important work, but that work is often limited to a specific function (surgical specialist, general practice, nursing, etc.) or a specific area (heart, vascular, gynecological, dermatological, etc.). Yet the medical profession, and the scientific research on which it is based, speaks to the public and is authoritative beyond the work of its individual specialized functions and jobs. It is a whole that is in fact larger than its parts because of the values that it (i.e., the medical profession) clearly represents in relation to the health and survival of humanity and the respect that its practitioners have by virtue of their publicly perceived embodiment of those values. Individual emergency management specialists, whether working in disaster response, predisaster preparedness, mitigation, or recovery operations, may be very talented. But the authoritative whole is missing to some extent.

It may be said that, with the advances made in hazard mitigation and the emphasis on hazard resilience that has emerged over the past two decades, the process of developing the authoritative whole (i.e., the profession and the professional voice of emergency management) became a possibility. It may even further be suggested that much of the ongoing discussion of a developing emergency management profession has benefited from these developments. But more is needed. The connectivity of emergency management more emphatically to sustainability, in addition to being a practical reality, offers a foundation for defining the fundamental value(s) to be served by an emergency management profession. It also offers, to the degree that its various specialties are all tied to that value, the prospect for a broader professional voice that may speak with authority and lend its perspective to all issues related to sustainability. As a sustainability profession, it has much more to say than any one of its specialists may say in their individual roles no matter how skillful they are.

Sustainability as a concept, its growing popularity in many quarters notwithstanding, is often a difficult value for many people to identify with. This is especially true if we think globally or in terms of temporal distances. When we talk about the problems of developing nations halfway around the world or the natural disasters that devastate them and their largely poor populations, we may feel sympathy, but we may also feel that it is not our immediate concern. After all, they are so far away, and what can we really do? When we hear of a climatological threat or natural resource shortage that may severely impact and harm future generations, again we may feel some general concern while justifying our inaction with the thought that we will all be dead when this happens. If not that, we often use the uncertainty principle (i.e., it may not be as bad as they say, there is no way of knowing for sure, etc.) to escape any sense of responsibility for immediate actions on our part. In other words, our sense of responsibility may be conditioned by proximity. Great distances, both spatial and temporal, sometimes (almost al-

ways actually) allow more immediate and short-term concerns to diminish any broader concern for global and long-term sustainability. But
spatial and temporal distance does not make a moral difference.

Obviously the quality (perhaps even the existence) of future lives depends on the world we leave in our wake. Future lives matter, and most
everyone would agree, but many question just who should be held responsible for them (in the communities in which we live or on a global scale). Often to ask the question is to begin the justification for not
holding oneself (or the present-day community) responsible. But not all
negative impacts associated with our short-term decision making are related to a place or time that is as far away as we think. Present lives and
the communities in which we live and work may also be endangered
by unsustainable practices. The spatial and temporal distances between
us and catastrophe may not be as great as we think. Within the span of
our lifetimes, decisions we make today may determine the number and
severity of natural and human-created disasters we will have tomorrow.
On a practical level, there are risks and vulnerabilities enough in the
short term that relate to unsustainable practices to make sustainable development and hazard resilience a practical as well as a moral priority.
Emergency management would do well to emphasize the practical nature of that priority in the short term while adhering to and promoting
long-term sustainability as a value.

It is one thing to claim that foresighted present actions and investments produce significant future benefits and another to support the
claim. Emergency managers and emergency management scholarship
must, of course, continue to build on and emphasize project-level cost-
benefit analysis that demonstrates a savings in recovery costs for every
dollar spent on hazard mitigation and that shows the other practical
benefits of forward thinking generally. Equally important, if not even
more, many documentable hazard risks are associated with the broader policy arena and the development decisions therein at all levels, national, state, and local. As a profession (i.e., a sustainability profession),
emergency management has a valuable perspective to offer with respect to the identification of risks, the assessment of vulnerabilities, and
the recommendation of steps that may contribute to sustainable and
hazard-resilient communities. This must become, to the degree that sustainable community development may actually be achievable, a more
important part of the public dialogue in relation to all relevant policy
decisions. This is to suggest that a forward-thinking or future-oriented

sense of ethical responsibility, as a correlate of power, is something that requires broad community involvement and applies to all public and private decision makers who influence the course and prospects for sustainable development. It is also to suggest a role for emergency management, as a sustainability profession, in providing its perspective on the identification and management of risks and vulnerabilities associated with public policies and private practices that touch directly on hazard resilience and sustainable community development.

Paraphrasing H. G. Wells ever so slightly, human history is a race between our knowledge and catastrophe. In emergency management, given what we have said about the necessity of knowledge (predictive and other) as a necessary ingredient in the anticipation of, response to, and prevention of disasters, the importance of this race is perfectly demonstrated. The running of a successful race also requires, as has been suggested, an ethical framework that is future-oriented. Knowledge, scientific facts, and the sound analysis of data are important. It tells us what is going on, and it may even help us to anticipate what will happen next, but it does not tell us what to do about it. What we should do depends on what we value and what we think about what we value. With respect to emergency management, what must be seen and understood to value is sustainable and hazard resilient communities. This is what tells emergency management what it should do with its accumulated knowledge and practical techniques. This is what tells the rest of the community what emergency management is and what emergency managers do when all is said and done.

The integrity of emergency management consists, one might suggest, in its doing what it should do. This is to say that the integrity of emergency management is tied to its focus on sustainability and its adherence to it as the value to be served by the application of its knowledge and the meeting of the forward-thinking responsibilities imposed by it. Such integrity without the necessary knowledge, knowledge especially relevant to emergency management as a sustainability profession, is indeed weak and useless. Likewise, the knowledge relevant to emergency management is without integrity if it is not tied to the development of hazard-resilient and sustainable communities. Such would indeed be dangerous and dreadful for the communities that emergency managers serve.

It should not take much imagination for emergency managers to comprehend that actions (individual and societal) we take today, no matter how beneficial in the short run, may result, in the long term, in predictable disasters and preventable tragedies. Avoiding that through the making of sustainable choices is generally a good thing. In fact, emergency managers who take hazard mitigation seriously, who work with others to create and maintain sustainable, hazard-resilient communities, are doing what they should do ethically speaking. That is, ideally, what defines them as professionals and defines their profession as well.

REFERENCES

AAEM. Alabama Association of Emergency Management's Administrative Policies and Procedures/Code of Ethics. Available at http://www.aaem.us/AAEM_FINAL.htm.

Bagby, L. M. (2002). *Political Thought.* Belmont: Wadsworth/Thomson Learning.

Beatley, T. (1985). Towards a Moral Philosophy of Natural Disaster Mitigation. *International Journal of Mass Emergencies and Disasters, 7*(1), 5–32.

Denhardt, R. B. (2003). *Public Administration: An Action Orientation.* Belmont: Wadsworth/Thomson Learning.

Dobel, J. P. (1999). *Public Integrity.* Baltimore: Johns Hopkins University Press.

Garofalo, C. (1999). *Ethics in the Public Service: The Moral Mind.* Washington, DC: Georgetown University Press.

IAEM. International Association of Emergency Managers Code of Ethics. Available at http://www.iaem.com/about/IAEMCodeofEthics.htm.

Jonas, H. (1984). *The Imperative of Responsibility.* Chicago: University of Chicago Press.

Lilla, M. (1981). Ethos, Ethics, and the Public Service. *The Public Interest, 63,* 3–17.

Locke, J. (1965). The Second Treatise. In Peter Laslett (Ed.), *Two Treatises of Government.* New York: New American Library.

Partridge, E. (1988). Ethical Issues in Emergency Management. In L. M. Comfort (Ed.), *Managing Disaster.* Durham, NC: Duke University Press.

Petak, W. J. (1985). Emergency Management: A Challenge for Public Administration. *Public Administration Review, 45*(1), 3–7.

Rohr, J. A. (1998). *Public Service, Ethics, and Constitutional Practice.* Lawrence: University of Kansas Press.

Schneider, R. O. (1993). The Ethical Dimensions of Emergency Management. *Southeastern Political Review, 21*(2), 251–267.

Schneider, R. O. (2000). Knowledge and Ethical Responsibility in Industrial Disasters. *International Journal of Disaster Prevention and Management, 9*(2), 98–105.

Schneider, R. O. (2002). Hazard Mitigation and Sustainable Community Development. *International Journal of Disaster Prevention and Management, 11*(2), 141–147.

Shue, H. (1980). *Basic Rights.* Princeton: Princeton University Press.

Shue, H. (1981). Exporting Hazards. *Ethics, 91*(4), 579–606.

Chapter 4

EMERGENCY MANAGEMENT, SUSTAINABILITY, AND CLIMATE: WARMING UP TO THE CHALLENGE

We have reached a point where we have a real emergency.

— Dr. James Hansen, NASA
Goddard Institute for Space Studies

We're in a giant car heading toward a brick wall and everyone's arguing over where they are going to sit.

— David Suzuki, Co-Founder of the David Suzuki Foundation and
an award-winning scientist, environmentalist, and broadcaster

INTRODUCTION

A recently published scientific study discussed evidence of the possibility for a major and irreversible shift in the global ecosystem. The approaching of a planetary-scale transition as a result of human influence seems more likely than we might wish to suspect. The growing plausibility of a planetary-scale tipping point, according to the study, makes it necessary to address the root causes of how humans are forcing biological changes (Barnosky et al., 2012). Naturally, this study was greeted with suspicion and dismissed in some quarters as another hoax perpetrated by scientists with an agenda. Unfortunately, that is the way many disturbing scientific findings are greeted. If they challenge the way we live or suggest that we must adjust our economic and material pursuits in a manner that requires us to defer short-term benefits, pleasures, or profits to manage a long-term threat, such studies are the target of unrelenting criticism. Even something as generally accepted and scientifically established in our minds today, such as the health risks associated with cigarette smoking or the dangers of second-hand smoke, had to endure the same reaction in the public, political, and economic realms. Time, and a growing body of scientific evidence, generally moves us to make reasonable policy and lifestyle accommodations to address the problem, but it never happens without a fierce fight.

The tipping-point study referenced herein is a paper published by twenty-two internationally known and respected scientists. It describes as urgent the need for better predictive models that are based on a detailed understanding of how the biosphere reacted in the distant past to rapidly changing conditions, including climate change and human population growth. This group of scientists from around the world is, in effect, warning us that population growth, widespread destruction of natural ecosystems, and climate change may be driving Earth toward an irreversible change in the biosphere, a planet-wide tipping point that would have destructive consequences absent adequate preparation and mitigation (Barnosky et al., 2012). This is fairly dramatic sounding stuff, but not even science can penetrate the unwillingness of the human mind to deal with the dramatic sounding stuff it does not want to hear, it does not want to believe, and it does not want to accept. Besides, science is often mistaken, right?

The suggestion that we may be looking at a planetary disaster within a few generations suggests, to many of the critics and skeptics of such research, that a scam to acquire more research funding is under way or the scientists in question have some sort of environmentalist or green political agenda. It rarely suggests, even to those who may be inclined to see the research and concerns raised as legitimate and important, the need to act proactively to anticipate and mitigate the potentially destructive consequences. Indeed, the mere suggestion of further study and, should that study confirm the threat, forward-looking action is frequently decried as an overreaction to "junk science" by the critics or an unnecessary degree of panic and/or just impractical by others given the uncertainties associated with the research, the extended temporal time frame, the shortage of resources to deal with it even if we even wanted to, and the always more pressing and immediate needs that require our attention. The options are thus reduced to ignore the science (e.g., it and the people who do it are untrustworthy because the scientists have an agenda) or ignore the threat (e.g., it's too far off, we can't be certain, we have other things to do first). Both reactions are often where the actual seeds of an ultimate disaster are firmly planted.

To be fair, many scientific prognostications are based on incomplete analysis or preliminary findings. This is why scientists keep doing what they do. They continue to observe and collect information. They continue to test the assumptions and models they have developed for explanatory and predictive purposes and to refine both them and our knowledge about what is to come and what its consequences will be for us. We would be foolish to act on every legitimate concern raised before we knew exactly what we were dealing with or what, exactly, we need to do. But even as the science advances and becomes more exact, its reliability having been rigorously demonstrated and proven in peer-reviewed work, there is a hard fight to be fought. Established political and economic interests are not easily swayed by even the best science, nor are we as individuals generally ready when confronted with sound and reliable science to refrain from our impulse to ignore either the science or the threat. It is as if there were a constant state of war between science and politics, between science and economic interests that feel threatened by it, between science and the natural human impulse to ignore either it or the unpleasant consequences it may predict. Nowhere is this state of war more in evidence in the United States in the early twenty-first century than over the science surrounding global climate change.

Global climate change is a topic over which many Americans disagree today but about which the science is growing more and more certain. In fact, the science is not much in dispute anymore. More and more people generally are, in fact, willing to admit that it is happening. But as much disagreement as ever seems to exist (much of it being promoted by self-interested parties who stand to profit by it) about its causes, likely impacts, and what if anything we should do about them. But it is indisputable that global climate change poses a challenge that will test the limits of nature, impact human populations, and influence markets and economies. The challenges of climate change are undoubtedly so far outside of our everyday experiences that we have difficulty knowing how to weigh them or how to evaluate the growing number of scientific studies that point them out to us. This difficulty is exasperated by the tension between science and politics over the subject as well as by that fact that scientific certainty is never absolute (i.e., uncertainties are a constant in science as in life itself). Nevertheless, it is increasingly obvious that we must accept the fact that global climate change is happening, and we must, on a number of levels, respond to it. It is also becoming, however inarticulately, clear that global climate change will have a profound but practical effect on and place new challenges before emergency management practitioners.

From an emergency manager's perspective, it is only necessary to accept that climate change is in fact happening and that it will have (is already having in fact) impacts on the communities in which he or she works and lives. This is not to diminish the importance of the body of work already done or the ongoing research into the broader causes and effects of climate change and its short- or long-term impacts. Rather, it is to suggest the practical need to take the fact of climate change into consideration even as that work unfolds to answer the larger questions and resolve any remaining uncertainties about the longer term projections. It has already been demonstrated in the reliable and accepted scientific work, which has led to a broad consensus, that we need to be prepared for the predictable impacts of a changing climate, and emergency management is an important part of that preparation. That is enough for the emergency manager to know and to begin with as she or he begins to contemplate the subject.

Just as we prepare for other disasters, natural and human-made, it has been argued that we should assess and prepare for the impacts of global climate change. It has been suggested that the process of preparing for, responding to, and adapting to regularly occurring natural disasters is similar to what is now required as we prepare for climate change. It has been argued, in fact. that the natural hazards literature can be heavily drawn on to guide governmental and community planning with regard to climate change preparedness (Tomkins and Hurlston, 2005). The same can be said with the application of adaptation and mitigation strategies (Thompson and Gaviria, 2004).

What is of special interest in the discussion of climate change is its inevitable practical implications for emergency management and its broader connection to the mission of creating and maintaining sustainable hazard-resilient communities. It is, despite the partisan disagreement it generates, the most important threat to sustainability generally and the most profound challenge for global decision makers in the twenty-first century. In addition to the imposed necessity of dealing with some of its predictable impacts, emergency management as a sustainability profession may have the potential to offer a unique and valuable perspective as a guide for policymakers who are striving to meet the challenges associated with climate change.

This chapter will begin with a discussion of what the scientific community agrees to be the general and predictable risks associated with climate change. We shall discuss these risks from an emergency management perspective and the practical need for the emergency management practitioner to factor them into his or her work. We shall take note of some basic strategies for adaptation, mitigation, and response to those challenges that by their definition must involve emergency management as an essential component. Most importantly, in relation to our purposes of further explicating the essential connection between emergency management and sustainability, we will discuss the viability of what will be called an emergency management perspective as a guide to policymakers at all levels as they seek to promote sustainability as they struggle to respond to an already changing climate. Let us begin with the general and predictable risks about which a scientific consensus may be said to exist.

WHAT WE KNOW: A NATURAL DISASTER SLOWLY UNFOLDING

One hears all too often climate change critics who assert that there is no scientific consensus about climate change. It is perhaps an understatement to say that the assertions of such critics are exaggerated (LePage, 2007). Climate change is not a new invention or an artificial concept. For more than a century, scientists have studied and documented it. It is a natural phenomenon, but it may also be driven by other factors associated with human activity on this planet. The science is sound. It does not eliminate all uncertainty, but even the best science cannot do that. The U.S. scientific community has long led the world in research on such areas as public health, environmental science, and issues affecting the quality of life. They have produced landmark studies on the dangers of DDT, tobacco smoke, acid rain, and global warming. But at the same time, a small yet potent subset of this community leads the world in vehement denial of these dangers. These merchants of doubt, in an excellent analysis by Naomi Oreskes and Erik Conway, are revealed to be a loose-knit group of high-level scientists and scientific advisers with deep connections to ideological political entities and to industry. Their work is an essential part of effective political and corporate campaigns to mislead the public and deny well-established scientific knowledge (Oreskes and Conway, 2010). Their primary mission is the creation of doubt to protect private economic interests or promote political agendas.

Despite the claims of the merchants of doubt and other critics, there is in fact an overwhelming consensus in the scientific community about global climate change and its causes. This is not to say that there is unanimity or certainty about the long-term impacts or longer-range assessments, but a baseline for analysis leads to agreement on various predictable impacts and the serious threats they pose. The major scientific societies (American Meteorological Society, American Geophysical Union, American Association for the Advancement of Science, Geological Society of America, American Chemical Society, U.S. National Academy of Science, and numerous international societies and academies) are in fundamental agreement that global climate change is happening, human activity is contributing to it, and the scientific understanding of it is sufficiently clear and compelling to require that reasonable steps be taken to respond to its threats (Union of Concerned Scientists, 2009).

It is well established that, no matter what we do today, the earth's average temperature will continue to rise. This is already something that is in motion and will continue even if we immediately engage in the best sustainable development practices and reduce greenhouse gas emissions (Haddow, 2009). A growing consensus has emerged and a compelling case has been made that this phenomenon will lead to (and in fact is already contributing to) predictable impacts. These include increasingly frequent and severe extreme weather events in the eastern parts of North and South America, Northern Europe, and Central Asia; more intense storms; higher global temperatures; increased glacial melting; and rising sea levels (Haddow, 2009).

To comprehend the significance of such changes, it may be instructive to consider what has long been known about the sensitivity of hurricanes to the conditions in the atmosphere and the oceans. It is well established that hurricanes are natural heat engines reliant on ocean warmth for their destructive power (Riehl, 1950). There is, within the scientific community, agreement that even slight increases in ocean temperatures will intensify the strength and quite possibly increase the frequency of hurricanes, especially in the Atlantic (Emanuel, 2001;

Holland, 1997; Knutson and Tuleya, 2004). While it may not be possible to attribute the occurrence of a hurricane to climate change alone, there is no doubting that climate change holds the potential to make the hurricanes that do occur much more dramatic and destructive. With the already experienced and the future projections for a continued rise in sea levels, the vulnerabilities of populations and infrastructures in coastal areas (e.g., higher storm surges, more severe flooding, etc.) will only increase (Mooney, 2007). Projections regarding sea level indicate that it will continue to rise between ten centimeters and a meter by the end of the twenty-first century (Houghton, 2004). But just like global surface temperatures, the rise in sea level will likely be greater and even quicker in some areas than others (e.g., Bangladesh is expected to experience a sea level rise of a full meter by 2050). The larger point to be made here is that even a modest rise in sea level can and will change the impact of regularly occurring tropical events in communities around the world and in the United States.

To compound the hurricane problem in the United States, most scientists agree that the impacts of climate change will be greatly enhanced by "the ever growing concentration of population and wealth in vulnerable coastal regions" (Mooney, 2007, p. 263). This means the threat imposed by global climate change in relation to hurricanes is heightened due to decisions we have made about coastal development. Demographic trends and development practices have set the United States up for rapidly escalating human and economic losses from hurricanes. Indeed, much of the increased hurricane-related damage experienced in the United States over the past two decades owes as much to government practices that subsidize risk and development strategies that invite it as it does to the forces of nature (Mooney, 2007). With the predictable consequences of a changing climate and with decades of unwise and unsustainable development practices, it is (according to many experts) safe to assume that we will have to work all the harder to ensure that our communities are resilient in the face of whatever the climate holds in store (Linden, 2006).

The merchants of doubt, some on the fringes of the scientific community and many outside of it altogether, deny the degree to which anthropogenic (i.e., human-caused) climate change is impacting temperatures or question how any changes in weather patterns (those already experienced and those projected for the future) are related to increases in greenhouse gas emissions. They either deny the negative environmental impacts of fossil burning fuels and other human activities or they underestimate their importance as factors contributing to the various threats to our sustainability (i.e., denial of the science or denial of the threat). They are unimpressed by the 2007 report issued by the Intergovernmental Panel on Climate Change in which more than 600 scientists from forty countries cited the growing evidence (peer-reviewed science) that human-made climate change (anthropogenic-caused global warming) is a reality that we must address (Broecker and Kunzig, 2008). They are unimpressed with and unfailingly critical of any new study that confirms this consensus within the scientific community. But the scientific community (97% of it) accepts with growing certainty that human activities have become a major causal variable of environmental change (American Meteorological Society, 2003). So too does the U.S. government, at least on some level.

The U.S. Climate Change Science Program undertakes research on behalf of thirteen federal agencies. Its 2006 report concluded that the observed patterns of climate change over the previous fifty-year period could not be explained by natural factors alone. The human production of greenhouse gases was a significant and serious contributing factor (Wigley et al., 2006). An increase in greenhouse gas concentrations in the atmosphere will, according to the U.S. Environmental Protection Agency, lead to an increase in the earth's average temperature. This in turn will influence the patterns and amounts of rain, reduce ice and snow cover, reduce the permafrost, contribute to rising sea levels, and increase the acidity of oceans. These changes will in turn impact our food supply, water resources, infrastructure, and health (EPA, 2012).

Climate change as we are already experiencing it and as we may reasonably anticipate some of its predictable future impacts poses identifiable short- and long-term challenges. But we should first say a word about some of the uncertainties with regard to longer range impacts. What we have been discussing thus far have been the observable and predictable slowly (and recently more quickly) developing changes that have reliably been associated with the changing climate. This is where the scientific consensus is strongest and the room for doubt is minimal (nonexistent actually). But what of the talk about some sort of worst-case scenario associated with abrupt climate change at some relatively distant future date, the reaching of tipping points, and the potential for a major and irreversible shift in the global ecosystem? Abrupt climate change refers to the crossing of a threshold that triggers or causes the climate to shift relatively swiftly from one state to another. Our state of knowledge is not yet sufficient to predict the timing of abrupt climate changes. The probability of an abrupt climate change may be low (Meehl and Working Group I of the Fourth Assessment Report of the Intergovernmental Panel on Climate Change, 2007), but it has happened in the past (Jansen and Working Group I of the Fourth Assessment Report of the Intergovernmental Panel on Climate Change, 2007). It is critically important to continue the study of such possibilities, whatever their probability. Indeed, their probability is subject to change (Barnosky et al., 2012), and that fact needs to be constantly analyzed. But one should not be so distracted by the debates about or the varying probabilities for abrupt change or worst-case (i.e., almost impossible to predict although quite possible tipping points) scenarios as to ignore the unwelcome climate changes that are occurring and those we can reasonably expect to occur (National Research Council, 2002).

With respect to climate changes we are already experiencing, there may also be some variance in shorter-term (say over the next fifty years) scenarios and projections. This is often pointed to by critics as proof that the science is not exact. That conclusion is, of course, to misinterpret the nature of the variations. These variations are the product not of inexact science but rather the differing assumptions fed into the analysis about the future choices we will make in relation to economic development, energy production, new technologies, and sustainability. The variations in the estimates of future sea-level rises, average temperatures, and so on have most to do with different projections about the amount of carbon dioxide in the atmosphere. Differing scenarios used to predict the future make, for purposes of analysis, differing assumptions about what we will do now, and they develop measures of how different decisions today may lead to varying (better or worse) outcomes tomorrow. Some scenarios make optimistic assumptions about our future choices, and some make more pessimistic assumptions about them. None of this, of course, should distract us from the consensus that climate change is real. It will and in fact already is altering the frequency and severity of weather extremes (i.e., heat waves, cold waves, tropical events, storms, flooding, droughts) and making the routine and regularly recurring natural hazards we must manage more dramatic and intense.

On a practical level, emergency managers might begin (whatever their take is on the scientific consensus regarding climate change) to see implications for the emergency management community. The expected effects of climate change, including predictable impacts on the frequency and intensity of future hurricanes, floods, wildfires, winter storms, heat waves, food- and water-borne diseases, will present them with significant changes related to the nature of the hazards already known and anticipated on a recurring basis in their communities. It may also present them with some new hazards as well.

It may be helpful, from an emergency management perspective, to define climate change as a natural disaster. It would also be important to remember that most disasters, as distinct from natural hazards, are in a significant proportion influenced or shaped by human causes (i.e., they are disasters by design). Thus, one must include as part of the definition of climate change as a natural disaster its relationship to the human decisions that determine the sustainability and hazard resilience of the communities it may impact.

Climate change may be viewed as a natural disaster already in progress. But it is different from any other natural disaster experience we have had. Unlike a tornado or a flood, climate change is not a short-term disaster with an easily identifiable and predictable short-term course that may be quickly or routinely anticipated, tracked, and readily amenable to immediate response and well-established procedures for recovery. It is happening or unfolding over an extended time period. It is moving in what one senses (inaccurately perhaps) to be slow motion when compared with every other natural disaster within our memory. It is both deceptively un-urgent at its onset and potentially more damaging that anything known through our previous natural disaster experiences with regularly occurring and expected events.

One must avoid the temptation, given the temporal distance between the present and the ultimate and most potentially devastating impacts of it, to view climate change as something far off. It must be seen and experienced as something already in motion. As such, it might be useful to think of climate change as a multifaceted, multi-event, prolonged, high-probability, high-impact global disaster. Its unfolding will span generations and constitute an ever present threat to be managed by human communities around the world. Its full magnitude is imperfectly known and in no small measure will be subject to the intelligent application or reckless avoidance of strategies to mitigate, adapt, or respond to it. This makes more urgent than ever the ongoing work to promote the development of sustainable communities and improve hazard resilience generally. No matter what we do or what we fail to do, the effects of a changing climate will impose global challenges to sustainability in general and to human safety and well-being in particular. From an emergency management perspective, what we must do (i.e., promote sustainable and resilient communities) is rooted in the practical knowledge and experience that tells us that there are clear tried-and-tested steps to be taken in the context of global climate change just as there are with any other natural disaster scenario.

EMERGENCY MANAGEMENT: A PERSPECTIVE TO GUIDE PRACTICAL ACTION AND POLICY

Emergency managers will have to deal with the short- and long-term impacts of climate change, including some of the more extreme effects that are possible. Yet it is probably safe to say that most are not fully aware of what this might entail. In addition to not wanting to become involved in a politically charged debate about climate change, they may also feel, to the extent that they acknowledge the importance of the threats it presents, that climate change is the responsibility of somebody else (i.e., it requires action far beyond anything in the practitioner's day-to-day environment and work).

There is no shortage of literature that discusses the link between climate change and risk assessment, risk management, disaster mitigation and prevention, and disaster response (Christoplos, 2008; O'Brien, O'Keefe, Rose, and Wisner, 2006; Schneider, 2011). Nevertheless, it is difficult for practicing emergency managers to translate this linkage into practical applications in their day-to-day work. The reasons for this are many. First, they may not have knowledge or a clear understanding of how climate change may relate to their day-to-day work. Second, often institutional roadblocks and jurisdictional barriers make it difficult to coordinate with the multiple entities and foster the wide-ranging relationships necessary to promote sustainable practices in relationship to climate change. Third, severe limits on time and resources may discourage them from broader efforts beyond their immediate tasks. Fourth, and finally, the emergency manager's traditional mindset (e.g., focus on extreme events in a shorter event horizon, preparedness for the next disaster, an all-hazards approach in which climate change would be one of many potential hazards and one easy to put off or delay specifically addressing given the more pressing concerns of short-term hazard potentials and resource limitations, and a general sense that the long-range thinking required for managing the effects of climate change is beyond their pay grade [and is ultimately the responsibility of others]) may make climate change less salient for them in their day-to-day work. Of course, many of these same justifications and rationales can be used to argue that the goals associated with sustainability are beyond the reach of the practicing emergency manager as well.

One particular aspect of climate change makes it even more difficult to incorporate into the emergency manager's day-to-day work. Preparedness and mitigation plans, for example, are based on hazard identification and vulnerability analysis for a specific community or geographical area. These, in turn, are based on historical experiences of previous hazards and disasters that may be expected to occur again. But climate change, as we have seen, holds the potential to alter the magnitude and frequency of extreme events that a community might reasonably expect to experience and present a few new or presently unanticipated events as well. Previous experience may not be an accurate guide for planning in this case. Investing time and resources in planning and decision making adapting to or mitigating climate change is further complicated by the variation in estimates of climate change impacts, intensity, and timing. Thus, it is easy to see how practicing emergency managers might regard climate change as a relatively low priority in their everyday work, one that can only distract them from the next flood, hurricane, or storm. Yet, as is the case with the development of sustainable and resilient communities in general, there is an important role for emergency management as a profession and as a community-wide endeavor (it takes a village, remember) and for the practitioner to play. To sort this out, it is first necessary to articulate what one might call the emergency management perspective on the problem of climate change. Secondly, it is necessary to see how the work of the practitioner (understanding its limitations and constraints) might incorporate that perspective in a practical way. Thirdly, and of perhaps greater importance than practitioners may know or scholars have previously addressed, it is necessary for the emergency management perspective to be integrated into the broader policy process and the decisions made therein. These three steps are in no small measure related to the development of an emergency management profession generally, and, to the degree they may be implemented, they broaden the responsibility for action beyond the often beleaguered individual practitioner.

An emergency management perspective, generically speaking, may be said to begin with an assessment of risks and vulnerabilities associated with hazard potentials. Such assessments in the practitioner's world are typically focused (as are preparedness, response, and mitigation plans) on a single jurisdiction and are limited by the standard parameters in which they are conducted. These parameters include a limited time frame of five to ten years and are tied to previous experience with recurring and predictable hazards within the immediate community or geographical area. Such parameters are of course inadequate in the face of potential climate change impacts that will span 50 to 100 years. Decades of mitigation and preparedness will be required to protect communities and populations. More importantly, the risk assessment perspective must expand or shift to a global perspective. The impacts associated with climate change, which will affect each specific community differently, will not respect political boundaries or jurisdictions. Likewise, and again speaking broadly, both risk/vulnerability assessments and the actions that will follow them will require the simultaneous involvement of many nations. All of this is to say that the context of climate change and the risks and vulnerabilities associated with it is indeed well beyond the experience and day-to-day world of the emergency management practitioner in a specific community. It means that risk and vulnerability assessments must be global and that national and international entities must provide mechanisms for integrating local planners and their assessments into a global network and be responsible for connecting local planners, and the specific disaster threats that will vary from location to location, to a broader nexus of climate change analysis and expertise.

Still thinking globally, it may be wise to place all risk assessments (local, regional, and national) pertaining to regularly occurring natural disasters into the context of climate change. This will in no way diminish traditional assessment, planning, or preparedness by any jurisdiction for its next regularly occurring or expected natural disaster, but it will connect that disaster to or assess it in the context of climate trends that are clearly observable and already changing the dimensions of regularly occurring and expected events. This would be critical to provide the necessary and valuable insight to be gained by taking each isolated disaster and placing it into the context of a global crisis unfolding over an extended period of time. Treating each individual natural disaster as a disaster within a broader unfolding disaster, climate change, suggests that each specific occurrence can be mined for information relevant to assessing the growing or shrinking risks and vulnerabilities that may be associated with the ongoing and developing phenomenon called global climate change. Finally, with respect to climate change, it is clear that the process of risk assessment must merge the expertise of scientists, emergency management specialists, political scientists, international relations experts, governmental entities at all levels, and relevant public sector entities.

What is being suggested, in relation to climate change as with any other issue related to sustainable development and hazard resilience, risk and vulnerability assessment (a basic function within the emergency management realm), is a broader responsibility related to sustainability, and it involves the entire village. The expertise of many, not just emergency managers, is required. But, and this is the important but, as a sustainability profession with expertise related to planning for hazard resilience and adapting and mitigating against various hazard threats, emergency management must be seen as having a professional perspective that may guide other experts connected to the assessment of risks and vulnerabilities just as it guides the work of a specific practitioner engaged in traditional risk/vulnerability assessments in a specific community. This is what one might refer to as the contribution of a profession as opposed to the work of an individual practitioner. The professional emergency management perspective thus, like the perspective of the medical profession broadly speaking that has conveyed the importance of preventive health care or the importance of diet, is an authoritative recommendation that guides the actions of all relevant actors. To the extent that what we are calling an emergency management perspective were in fact clearly articulated, it would offer an invaluable guide to policymakers and private sector leaders who must incorporate thinking about climate change-related risks and vulnerabilities into their decision making.

Continuing the broader emergency management perspective, the assessment of risks and vulnerabilities is followed by planning and preparedness activities. Preparedness involves, of course, the creation of capacity (technology and personnel), expertise (training), and resources to respond effectively in the interest of public safety and the protection of community infrastructure in the face of a disaster event. The planning and preparedness function and its necessity as a component in any hazard-resilient community is well understood and implemented by practitioners in their specific communities as they plan and prepare for the next anticipated or predictable regularly occurring event in their jurisdiction. But just as the global nature and the extended temporal framework of climate change requires risk and vulnerability assessments on a broader scale, well beyond the individual practitioner at the community level, so too preparedness responsibilities require broader involvement. But the emergency management perspective can guide that effort generally.

It is a common practice in emergency management to prepare for the worst and hope for the best. In the case of global climate change, preparing for the worst takes us to genuinely uncharted territory. Rising sea levels in a worst-case scenario, for example, could inundate lowlands and shore areas in places like Bangladesh, The Netherlands, New York City, New Orleans, and Florida. Almost half of the people alive today live in coastal zones that would be subject to flooding and perhaps even significant loss of habitable land with even small increases in global sea levels (Houghton, 2004). Populations living in some of these areas might have to out-migrate. Resettling large migrating populations peacefully and efficiently is a challenge for which no plan currently exists. Such a possible catastrophe means that the worst-case response scenario must expand beyond thinking in terms of single jurisdictions and think in terms of entire civilizations. Even assuming the best, we know that the observable effects of climate change and human land use policy have already combined and resulted in the loss of vast tracks of previously arable land. This makes the issue of relocating populations a bit more real and more immediate perhaps than a worst-case scenario that might easily be considered so unlikely as to be ignored by too many people (including policymakers) who should, but don't, know better (Bissell et al., 2009).

Thinking beyond unprecedented situations and the capacities that may be required for them, the expected intensification and increased frequency of storms, floods, droughts, wildfires, and the already changing and more dramatic impacts of other normal and recurring events means the increased possibility for mega-disasters that may overwhelm existing local and regional response capacities. Both an increasing number of events and an intensification of their impacts can drain budgets more quickly than anticipated as well as present damage on a scale not previously experienced or presently anticipated. As such threats may be predicted, documented, and multiply as the intensification of the expected becomes a new normal that exceeds our previous experiences, and as impacts affect broader areas cutting across jurisdictional lines, resources for preparedness and recovery will have to increase significantly to facilitate effective response and effective recovery and to enhance future hazard resilience. This thought is unsettling given the worldwide debt crisis, the resort to austerity as prime response to it, and the inability of governments everywhere to maintain even their basic level of services.

Whatever the future may hold, it is clear that the larger and unfolding disaster that is climate change will make natural disaster preparedness more critical, and it will require more resources globally than currently seem to be available. It will also, given the possible magnitude of its impacts, require the increased capacity to combine or integrate efforts across jurisdictional lines, even internationally where necessary, to respond to mega-disasters and worst-case scenarios. Of course, all of this is well beyond the control and perhaps even the concern of the individual practitioner in a specific community. But it is a critical concern of the emergency management profession, which should, ideally, have something to say about it to national and international policymakers who will make the decisions today that will determine the level of preparedness for some of the dramatic tomorrows that are predictable on the near horizon and the more dramatic days after these tomorrows that are being projected by some reliable science on the farther horizon.

Continuing with our emergency management perspective, hazard mitigation is the phase of emergency management that emphasizes the taking of proactive steps to prevent disasters, reduce their negative impacts, or adapt to hazard potentials. As discussed in Chapter 2, it typically takes place before anticipated disaster occurrence or in the immediate aftermath of a disaster that may be expected to occur on a recurring basis. Mitigation, like every other disaster phase, is thought of first as a local function, and it is, in fact, most successful when conducted at the community level. But climate change, the global hazard threat, will require unprecedented action across jurisdictions. Mitigation and adaptation to it requires action once again well beyond the scope of the individual practitioner at the local level. Again, emergency management as a putative profession should have something to say about that.

With respect to climate change, the scientific and policy communities have already emphasized mitigation in the form of reducing the amount of carbon and other greenhouse gases in the atmosphere. The reduction of greenhouse gases (GHGs) is really the only purely effective mitigation measure. The rest of what we will discuss as mitigation refers to adapting to climate change and increasing the resilience of communities. Adaptation and mitigation are part of the general mitigation function in emergency management. Often the terms are used interchangeably, but they are distinctly different in practice.

The efforts within the international community to cooperate in the reduction of GHG emissions are ongoing although with far from perfect results (PEW Environmental Group, 2008). While progress is being made in small bits and pieces, it is accurate to say that the relative complexity and political challenges of reaching and implementing agreements to reduce GHG emissions combined with the fact that such agreements represent a long-term effort that will take decades to achieve means that reliance on reductions in GHG emissions alone is not nearly enough. The climate is already changing. Its effects are already being felt. More immediate steps need to be taken to provide short-term relief and to set in motion policies and actions that will yield significant long-term benefits. This can best be done by devising and implementing adaptive strategies that contribute to the building of more resilient and sustainable communities.

Hazard mitigation that takes the form of advanced action to reduce the impact of expected or predicted events, reduces the risks to human life and property, and protects community infrastructure and vitality requires as a first priority the development of sustainable communities (Schneider, 2006). Sustainability in this context, and as stated in Chapter 2, means that a community can tolerate and overcome damage, diminished productivity, and negative impacts on the quality of life that may result from a disaster occurrence (Mileti, 1999). Let us briefly review some of our discussion of this from Chapter 2.

Two basic types of hazard mitigation may contribute to hazard resilience. These, as noted in Chapter 2, consist of structural and nonstructural techniques that are routinely employed by local communities as they seek to build hazard resilience into their planning for sustainable community development. Structural mitigation (also called hard mitigation) includes the strengthening of buildings and physical infrastructure exposed to hazards. Reducing the vulnerability of infrastructure to disaster damage is accomplished through a variety of techniques, including building codes, improved engineering and design, improved construction codes, and improved technologies. Its purpose is to increase resilience or damage resistance. Nonstructural mitigation (also called soft mitigation) emphasizes directing development away from known hazard or high-risk locations, relocating to safer locations developments that experienced repeated damage, and maintaining the protective features of the natural environment that may absorb hazard impacts and reduce their damage to human populations and community infrastructure. Nonstructural mitigation also includes the integration of all decisions relevant to economic development, structural development, energy production and consumption, the environment, and the needs of at-risk populations as necessary components to reduce hazard risks and promote community sustainability. Even within a specific community, hazard mitigation considered thusly and connected to sustainable development requires the bundling of efforts (it takes a village, we must keep saying) of an entire community. But the emergency manager is a key player at the local community level and, while not able to create hazard-resilient sustainable community alone, a necessary part of the community development network.

Given what we know about global climate change, communities around the world must be prepared to adapt to more frequent and possibly more destructive disaster events. Communities will have varying and different levels of climate-induced risks to be sure, but all will have risks that become chronic as time goes by. Hazard mitigation techniques that are already employed successfully can teach us much, and they are of course an important building block. However, two things must be noted that might give us pause. First, governments around the world and at all levels have taken an inconsistent approach to dealing with regularly occurring and expected natural disasters. Second, governments have been even more inconsistent in addressing the impact of climate change. It is also worth noting that, with respect to hazard mitigation, climate change is only beginning to be considered as a planning variable, and such consideration as it does receive is not consistent across jurisdictions.

Policymakers at all levels must promote the necessary connection of traditional hazard mitigation to efforts to measure and anticipate the future impacts of climate change. Indeed, the basic knowledge we already have about climate change (long-term uncertainties notwithstanding) should be a strong call to enhance or improve mitigation planning. The fact that climate change may increase the frequency and intensity of expected and regularly recurring events, changing their character and their future impact potential, should provide renewed impetus for mitigation efforts in every community around the globe.

Moving from the emergency management perspective to the practical world of the practitioner at the community level, it should first be noted that the discussion of the relationship between community-based efforts to develop mitigation and adaptation strategies for climate change and its linkage to emergency management is in its infancy, but it is happening. Emergency managers are coming to be regarded as necessary stakeholders and participants, even if not leaders, in such efforts. It is becoming apparent as communities have begun to dialogue and plan in regard to it that local adaptation strategies to climate change are clearly linked to emergency management concerns, especially with respect to disaster risk reduction and hazard resilience. Recognizing this linkage, the province of Ontario, Canada, for example, included emergency management in its planning process for responding to climate change, saying that it (i.e., climate change) was also a critical variable to be factored into the ongoing development of its community emergency management program (Hyslop, 2011). An awareness of this linkage has also encouraged many others to specifically mention the inclusion of emergency management in community-based planning groups working on climate change adaptation strategies (Labadie, 2011). There is no disputing the value of including the emergency manager and the emergency management perspective in such planning activity at all community, regional, and even national levels. But beyond such participation, what might the practicing emergency manager incorporate into the day-to-day work he or she performs that would relate to this stated linkage between climate change adaptation and emergency management?

To begin with, emergency managers are constantly anticipating and dealing with the impact, location, frequency, and occurrence of natural hazards such as hurricanes, wildfires, floods, tornados, and winter storms. But as we have seen, climate change may well change the impact, location, frequency, and occurrence of these events. As such, the historical data that are typically the basis for risk identification and vulnerability assessment by emergency management agencies and practitioners may not accurately predict future events. Emergency managers, as a practical matter, will need to begin understanding how climate change may relate to their typical tasks. Climate change projections relevant to their specific communities will affect their identification and selection of hazard mitigation strategies, the preparedness activities that their jurisdictions will need to undertake, the nature and scope of their response operations, and the identification and implementation of recovery strategies.

The Center for Naval Analysis (CNA), a nonprofit research organization, has conducted some research that may help decision makers and emergency management practitioners begin to integrate climate change into their standard operations (Silverman, Levy, Myrus, Koch, and DeGroot, 2010). This research is based on the identification of natural hazards that are of significance to emergency managers in various parts of the United States (e.g., hurricanes, floods, wild fires, winter storms, heat waves, flooding, etc.) and identifies the expected impact of climate change on them. One should note that many states and local governments, as part of their efforts to address the issue of climate change, have also gathered or assembled research on the potential impacts of climate change on the regularly recurring hazards within a given jurisdiction. An examination of such research leads to the conclusion that climate change impacts in the United States are identifiable and need to be factored into our everyday thinking on a number of practical levels (Karl, Melillo, and Peterson, 2009). This includes the integration of climate change projections into emergency management and preparedness policy and has implications for each of the disaster phases with which the emergency manager is already familiar and involved (Silverman et al., 2010). The CNA report makes several practical suggestions that relate to each of the disaster phases.

With respect to the **mitigation phase** of emergency management, mitigation strategies will have to adapt to the anticipated impacts (short and long term) of climate change. Coastal communities, for example, may be faced with more frequent and severe tropical events in the next ten to forty years. This means there may be a need for changes in strategies and policies related to land use planning, zoning regulations, and the like. There may also be implications for coastal wetlands rehabilitation, flood plain management, business interruption insurance, homeowners insurance, and so on. Severe winter storms, heat waves, wild fires, and tropical events may also impact the transportation and energy infrastructures and require considerable resources for road repair and power restoration. The point is, such reasonably expected impacts related to climate change need to be factored into state and local mitigation planning efforts that are conducted by emergency management agencies (Silverman et al., 2010). The rigorous risk assessments that are part of the emergency management realm might, as adapting to climate change becomes a part of them, suggest the utility of proactive engagement by emergency management practitioners with state and local climate research groups to secure data to support more accurate forecasts of climate change affects. Coordination with local, regional, and state climate change adaptation planning groups would also be advised to support hazard identification and risk assessment. Suffice it to say, hazard mitigation must integrate climate change into the mix of factors that it considers. This applies to all actors, including emergency managers, who perform tasks and make decisions relevant to the sustainability and hazard resilience of each community.

The disaster **preparedness phase** must take climate change into consideration as well. Preparedness activities will need to account for the changing risk profiles of communities and their populations associated with climate change. Planning assumptions and scenarios must be reexamined to address the increased frequency and severity of the natural hazards that can be expected to occur within a specific community. Changes in risk profiles may require reassessment of capabilities and a determination about which (if any) additional resources are required for preparedness efforts. Changes in risk profiles may also have implications for new or different impacts on vulnerable populations that need to be considered. Disaster preparedness, like mitigation, might benefit greatly from the inclusion of the emergency management community in local, regional, and state climate change adaptation planning groups (Silverman et al., 2010).

The **response phase** will be more complex due to climate change impacts for a number of reasons. First, local resources may be overwhelmed by more frequent and severe hazards. As we have suggested earlier in this chapter, the magnitude of impacts may require more complex collaboration in response efforts across jurisdictions. Mutual aid among agencies within a community may not, in the face of more frequent events and an escalation in their severity, be able to meet the challenge. Operationally, disaster response may more routinely require more help from a wider array of state and federal agencies. Climate change impacts will require both a better assessment of larger and more frequent response efforts on local budgets and the anticipation of some command and control challenges associated with larger scale events that are likely to become more common (Silverman, Levy, Myrus, Koch, and DeGroot, 2010). New strategies will be required for managing more frequent and complex disasters.

The **recovery phase** will also bring new challenges related to climate change. The efficiency of the recovery process may be severely tested as it will have to deal with more frequent and more costly disasters. Recovery costs may escalate considerably. The changing risk profile and the increased frequency and severity of various natural hazards will also require emergency management officials and policymakers to make different and more difficult decisions about when to rebuild in certain areas or when to relocate. The question of how to rebuild may also require a different answer as the rebuilt or new construction will have to be stronger and more resilient. On a practical level, the creation of decision criteria to guide these decisions in relation to projected climate change impacts is likely to be needed (Silverman et al., 2010).

From a practical point of view, the following seems reasonable. The practicing emergency manager may not yet regard climate change as a concern he or she needs to take seriously, but it is clear that climate change will have a serious effect on the work he or she does. This being the case, it is not unreasonable to suggest that its impacts must, where possible, be integrated into the thinking, planning, and implementing of the practical work of emergency managers in each disaster phase. Whatever their previous inclinations or their perceived limitations of their relationship to climate change, it is clear that emergency managers must warm up to it. Dealing with the consequences of climate change will of necessity involve them and the work they do in communities across the country and around the world. Integrating it and its impacts into their work is both possible and necessary from a practical point of view. But clearly, there are other actors on a broader stage and the practicing emergency manager's work can be successful only to the degree that these other actors are engaged. Thus, some broader policy concerns must be addressed.

Policymakers at all levels can (must) work to make climate change preparedness, response, and mitigation more consistent and effective. This requires that policymakers make it a priority to study and understand climate change, its projected impacts, and to responsibly improve all efforts to address the threats and risks it poses on populations and communities. Deciding what is necessary and feasible requires rising above partisanship and self-interest to develop a common or shared perspective for policymaking that will help to produce the consistency necessary for progress in mitigating and adapting to climate change. An emergency management perspective, to the extent that one may be said to exist, may be the best vehicle for promoting consistency and progress. A consensus of perspective, so to speak, is necessary at all levels (local, state, regional, national, international) to promote actions, both local and across jurisdictional lines, that are necessary for success.

Putting an emergency management perspective on climate change into practice at the policymaking level means ending all counterproductive and partisan debates about whether climate change is real or imagined. That discussion does not merit another breath of precious life or a moment of our valuable time. Likewise there should be no debate about the need to act in response to the evidence that demonstrates the expected impacts of climate change. There is much for policymakers to debate and about which to disagree, but not the central reality that climate change is happening and the growing certainty that we must respond to it. Putting an emergency management perspective into practice as a guide to policy means focusing on the scientific evidence, the preponderance of which (i.e., almost all of it in fact) is in strong agreement with the need to take some form of intelligent action to manage and adapt to the effects of a changing climate. It means constant refinement and improvements of all efforts to accurately monitor a changing climate and to better understand its potential impacts. It means targeting for assessment the risks and vulnerabilities that can reasonably be identified on the basis of reliable scientific consensus. It means above all else a process that is analytical, fluid, and proactive. Policymakers at all levels should promote policies that pursue six basic objectives.

Objective number one is to *assess climate change impacts in all sectors.* The emphasis should be on assembling the best scientific information available. As this is done, it should be remembered that while sufficient information may not be conclusively available with respect to some long-term projections, this does not negate the need to act in the short term or justify the dismissal of short-term projections and the identification of definite and accurate trends already observable and about which there is an indisputable scientific consensus.

The second objective of climate change policymaking should be to *promote and support hazard identification and vulnerability studies in relation to the projected impacts of climate change.* These will vary from community to community and region to region, but all will be affected. Understanding the varying impacts on different communities and regions is essential, especially with respect to the anticipation of future hazard occurrence (its frequency and severity) and its likely variance from what previous disaster experience may lead a community to expect normally. Normal is changing in every community, and it is important to factor the new normal defined by climate change impacts into risk assessments.

Objective number three relates to the actual implementation of mitigation and adaption strategies. Policymakers should make it a priority to *identify and implement actions to mitigate identified risks and vulnerabilities as well as promote adaptive strategies that contribute to hazard resilience.* This includes the reduction of greenhouse gas emissions as a mitigation action, protections of the natural environment, and adaptations in the constructed environment that create resilience to hazards, reduce their impacts, and contribute to sustainability in the face of changing threat assessments. This should, ideally of course, complement and enhance the work already ongoing to assess risks and vulnerabilities and to prepare for natural disasters.

A fourth policy objective that makes great sense, given what we already know about climate change and the frequency and/or severity of already recurring disasters, addresses the need to *improve response capabilities for all natural disasters and expand response resources to meet the changing risk profiles associated with climate change impacts.* It will also be critical, in relation to this objective, to promote some planning for unprecedented mega-disasters and worst-case scenarios. These in particular will exceed current coping strategies and response capabilities.

The fifth policy objective should be to *promote regional, national, and international cooperation in assessing risks and vulnerabilities and in responding to increasingly complex and even unprecedented mega-disaster scenarios.* The development of greater capacities for cooperation across jurisdictional lines to respond to natural disasters has never been more urgent, and this need will only increase as the effects of climate change continue to unfold.

Finally, policymakers must promote and support ongoing efforts to *measure progress and update all assessments of climate change impacts.* The situation is fluid. Events will unfold. Scientific information will continue to expand and with it the knowledge base for intelligent policy. Hence, policy must be nimble enough to adjust to changing circumstances and new information.

An emergency management policy perspective, should something like it develop along the lines here suggested, might enable policymakers to speak the same language, participate in the same analysis, and pursue the same goals. It does not mean they will easily agree on the preferred means or the best policies to meet those goals, and there will be many partisans and competing interests to continue an animated debate. But an emergency management perspective elevates the conversation just enough perhaps to make the resulting policy outcomes more consistent and responsible in addressing the global challenge known as climate change. It would also no doubt contribute a sense of direction, perhaps even a few more resources committed to addressing climate change, and make easier the integration of climate change considerations into the preparedness, mitigation, response, and recovery efforts of emergency management as well and contribute to the mainstreaming of emergency management into the broader climate adaptation planning efforts of the communities they serve.

THE FUTURE OF EMERGENCY MANAGEMENT

Emergency managers *sometimes* ask, how bad is bad? Worst-case scenarios suggest that the impacts of warming temperatures, heat waves, heavy rains, droughts, stronger storms, sea-level rise, and all of the other effects associated with climate change could make bad very bad indeed. Some places, even densely populated places, will become uninhabitable and others perhaps a bit more habitable. Deserts could span across the American Northwest, and the Mediterranean might have the climate of North Africa. Sea-level rise will subject coastal areas, and almost half of the people on the planet today live in coastal zones, to stronger storm surges and much more severe flooding. Some scenarios suggest serious water shortages and the loss or reduction of crops due to increasing temperatures. The worst scenarios suggest a future with hundreds of millions of people – displaced, thirsty, and hungry – seeking to escape not just sea-level rises but moving away from scorched croplands and waterless wells (Houghton, 2004; Lynas, 2007). All of this will be accompanied by weather extremes that will be intense, increasingly destructive, and ever more costly. Of course, such projections are less than persuasive to almost everyone if you are honest about it.

Projections about how bad things could be are easily dismissed because they refer to cumulative effects that are hard to predict with absolute precision and will be the product of both natural events and human decisions unfolding over many decades. Add to the mix the heavy dose of the predictable skepticism that meets any attempt at alerting us to an almost unimaginable and far-fetched future global disaster that requires dramatic action in the present, add the investment in the status quo by well-financed interests that have an immediate stake in promoting short-term profits and the unsustainable practices that will produce both them *and* the long-term global disaster, and add the general human tendency to not be future-oriented, and the answer to the question of how bad can bad really get seems rhetorical and insignificant in the immediate scheme of things. Scientifically sound and well-documented projections about how bad things might get certainly will not, even where science might reasonably dictate that certain forward-looking actions are prudent, inspire action. Debate yes, but action not very likely. With respect to emergency management in particular, the answers to other questions that are more immediate to the experience and concerns of practicing emergency managers will probably inspire action.

Emergency managers *always* ask what it is that we need to do now. Where should our resources and efforts be concentrated? With regard to climate change, whatever their beliefs about the science, the long-term impacts, and the possibility of any worst-case scenario, they first and foremost want to know three things. First, they want to know whether this is something they really need to deal with; second, they want to know what specifically they must deal with and when they will need to deal with it; and third, they want to know how what they must deal with fits in with everything else they have to deal with in their work. Much of our discussion in this chapter has answered these questions.

Emergency managers, regardless of whether they are fully aware of it, are already dealing with global climate change. It is already influencing and altering the dynamics and nature (i.e., frequency, intensity, severity, etc.) of regularly recurring events in all jurisdictions. The number of record heat waves, the growing scope and intensity of wildfires, the frequency and intensity of storms, and the damages being caused by these recurring natural events are already and quite noticeably impacting the day-to-day work of emergency managers. Resources for response are struggling to keep pace with events, the costs associated with recovery are increasing, and the implications for predisaster planning and hazard mitigation are increasingly serious in relation to the impacts of an already changing climate. Is climate change something emergency managers should deal with? Well, they are already dealing with it, and in the normal course of their work, they will be required to deal with it all the more as it alters the course of the recurring hazards with which they are already dealing in their jurisdictions.

The answers to the questions of what specifically emergency managers must do and how it fits into the rest of their work depends on how, one would suppose, climate change will be perceived as relating to and/or impacting the work that emergency managers already do. Is climate change a wholly new hazard threat that requires a unique set of responses or does it only make existing hazards worse (i.e., severity, duration, geographic spread), thus not requiring special or novel adaptations for the emergency manager? The analysis in this chapter would lean to the former to some extent (i.e., it requires a unique set of responses). To the degree that climate change is a long-term factor, to the degree that it will alter the regularly occurring events within a jurisdiction (i.e., the extreme and previously perceived unlikely become more normal and commonplace), it is clear that standard thinking and some standard procedures will be inadequate. For example, as climate change alters the frequency and severity of events, coping and response mechanisms and all disaster-related planning based on past vulnerabilities (i.e., historical experience) will be insufficient to anticipate current and future vulnerabilities.

As noted in our earlier discussion, each of the disaster phases (i.e., mitigation, preparedness, response, and recovery) that occupy the time and attention of emergency managers in their normal work will need to engage in some forward thinking that incorporates climate change information. The changes in risk and vulnerability profiles, the changing frequency and increased severity of weather events, and the escalating costs associated with them that are linked to climate change all must be factored into the normal workload. This requires not the performance of new tasks so much as the connecting of the tasks already being performed to new information and expertise not typically associated with emergency management functions. But just as the emergency manager's role in contributing to sustainable and hazard-resilient communities in general requires their efforts to be connected to broader community development decision making, so too their efforts must be connected to all planning at the community, regional, and national levels to adapt to a changing climate. Thus, as previously noted, emergency managers must have a seat at the table in every community that is planning for and adapting to climate change. Equally important, emergency managers must seek out and incorporate into their normal operations all scientific and predictive information relevant to maintaining the sustainability and resilience of the communities they serve in the context of its predictable natural hazard threats. Most states and many local communities have begun planning for the predictable impacts of climate change, and a great deal of information regarding these impacts is in fact available.

Given the nature of their work and the potential impact that climate change may have on it, emergency managers need to be involved. The strategies needed to adapt to climate change will require approaches that not only enhance environmental quality but, given the nature of the threats posed and some of the more extreme events already being experienced, also identify and implement plans that combine disaster risk and vulnerability reduction, pre-disaster preparedness, post-disaster recovery, with the broader planning for environmental and community sustainability. The emergency manager's expertise will be needed as part of this discussion. It is critical that emergency managers take an active part in all aspects of planning and implementing activity related to climate change adaptation. It is also critical that they find ways to enhance the effectiveness of their traditional functions by incorporating climate change adaptation into their everyday work. This is another instance, like the broader discussion of sustainable development and hazard resilience as a core component in it (see Chapter 2), where emergency management perspectives and skills can and should be mainstreamed into overall community planning and development. But emergency management as a profession, beyond the specific work of its practitioners, may have a role to play that is even more important.

To the degree that it may indeed be (or more accurately become) a sustainability profession, emergency management must have a voice that informs the thinking of policymakers generally and that has something important to say in the broader discussion of risk and vulnerability assessments and the strategies proposed to manage them as it relates to all climate-related issues touching on the various threats to sustainability. Global climate change impacts the work of the emergency manager, it requires the emergency manager to play a role in the necessary efforts to adapt to it, and it cannot be addressed without addressing at the policymaking level the concerns they deal with on a regular basis (risk and vulnerability reduction, disaster preparedness, response, etc.). What one might call the emergency management perspective with its focus on forward thinking to anticipate, mitigate, and respond to threats to sustainability and resilience posed by all hazards has broader applications than one might typically think. In discussing the role of such a perspective with regard to planning for and maintaining sustainable communities in Chapter 2, it was noted that hazard mitigation was a function that tied emergency management directly and unavoidably to all community planning for social and economic development as well as to environmental well-being. We noted that community planning for sustainability could not meet its goals if it did not ensure that economic and political decision makers operate with a full awareness of the risks to people and property vulnerable to natural and other hazards. To be sustainable, communities must make it a priority to anticipate and find solutions to the risks and vulnerabilities associated with hazards. That includes the risks and vulnerabilities associated with climate change.

Emergency management scholarship, combining with and incorporating the relevant scholarship of many disciplines, must emphasize as never before the need to anticipate and find solutions for hazard risks and vulnerabilities. Global climate change represents that need at its most urgent level and demonstrates the most powerful obstacles to meeting it. Climate change is perhaps best seen, as previously stated, as a multifaceted, multi-event, prolonged, high-probability, high-impact global disaster. This is to say, from an emergency management perspective, we have reached the point where it has already become a real emergency. It is, to be sure, not perceived that way by many. This is to be expected given the extended temporal distance that exists between some of the anticipated impacts that can be projected and their inevitable arrival. In fairness, it must also be admitted that the precise level and seriousness of some of these impacts is difficult to pinpoint with absolute certainty. This gives plenty of room for doubt, denial, or simply the impulse not to react until it all becomes more immediate. But the science is increasingly clear, and recent experiences have begun to confirm that it is serious and requires our immediate attention.

As anyone familiar with the emergency management literature knows, most people do not think about disasters until they happen. They underestimate the potential for a disaster or, given the statistical odds, roll the dice thinking it won't happen to them or their community. They move from either denial of the threat or avoidance of it straight to the despair that accompanies a disaster occurrence. To their credit, many pick up the pieces, rebuild their lives, and persevere, but the cost of all that is much higher than it needs to be, and many of the losses were entirely predictable and avoidable. Emergency managers know there is something in between denial or avoidance and despair and rebuild that can alter events. That intermediate something is the reality labeled "we can make a difference." This intermediate step defines the role of emergency management most broadly. This role needs to be played not only at the level the trade is plied but by an emergency management profession that lends its perspective and methods, combines its expertise with the expertise of others, builds the case for the sort of forward thinking that can make a difference by reducing risks and harms, and influences decision makers both public and private to adapt an emergency management perspective to their work.

Global climate change is, with respect to the reliable scientific foundation that identifies its already felt impacts and its reasonably projected threats, the number one long-term challenge to be addressed with respect to both sustainability and hazard resilience. This makes it a priority for every community. This makes it a priority for emergency management. It makes sustainability perhaps the most important word of the twenty-first century, for the threats to it resident in climate change hold the potential for unprecedented disaster. This sort of talk makes many people uncomfortable, but it is the sort of talk that is far more important than those who dismiss it may ever know, at least until they skip straight from denial to despair. And let's be clear, it is easy to dismiss such talk given the lack of total certainty regarding longer range projections and the extended time involved before many more serious impacts will be felt. It seems neither real nor immediate enough to warrant such talk. But an honest and informed look at what science has reliably shown us and the growing number of weather extremes we are beginning to experience are more than enough to make such talk reasonable to prudent and concerned people.

Unlike many disasters that humanity denies or avoids until they happen and greets with despair when they arrive, disasters which people can rebuild after and recover from whatever the costs and however avoidable they might have been with forward thinking and planning, climate change has the potential to be a disaster from which denial and/or avoidance may lead to a despair from which there is no recovery, at least for many. It may be that we are heading for a brick wall. Denial and/or avoidance may guarantee some of us a more comfortable seat on the planet when it hits that wall, but it won't spare us the damage or enable us to survive the wreck. This may be the one time where lack of forward thinking, perpetuation of unsustainable practices, and failure to adapt and mitigate may produce a disaster from which we will never recover. Emergency management, both in the work of its practitioners and as a profession, must warm to the challenge of climate change and be an integral part of preventing that disaster. That is the essence of emergency management and its central role in the creation and maintenance of sustainable and resilient communities. It may also be the ultimate measure by which the success or failure of both emergency management and humankind will be judged in the end.

REFERENCES

American Meteorological Association. (2003). Climate Change Research: Issues for the Atmospheric and Related Sciences. *Bulletin of the American Meteorological Society, 84.*

Bissell, R. A., Bumback, A., Levy, M., & Echebi, P. (2009). Long-Term Global Threat Assessment: Challenging New Roles for Emergency Manager. *Journal of Emergency Management, 7*(1), 18–38.

Broecker, W. S., & Kunzig, R. (2008). *Fixing Climate.* New York: Hill & Wang.

Barnosky, A. D., Hadly, E. A., Bascompte, J., Berlow, E. L., Brown, J. H., Fortelius, M., Getz, W. M., Harte, J., Marquet, P. A., Martinez, N. D., Mooers, A., Roopnarine, P., Vermeij, G., Williams, J. W., Gillespie, R., Kitzes, J., Marshall, C., Matzke, N., Mindell, D. P., Revilla, E., & Smith, A. B. (2012). Approaching a State Shift in Earth's Biosphere. *Nature, 486,* 52–58.

Christoplos, I. (2008). Incentives and Constraints to Climate Change Adaptation and Disaster Risk Reduction: A Local Perspective. The Commission on Climate Change and Development, Stockholm, Sweden. Available at http://www.ccd-commission.org/Filer/pdf/pb_incentives_linking_climate_change.pdf. Accessed June 20, 2012.

Emanuel, K. (2001). The Contributions of Cyclones to the Oceans' Meridional Heat Transport. *Journal of Geophysical Research, 106* (14), 771–781.

EPA. (2012). Climate Change Impacts and Adapting to Change, Environmental Protection Agency. Available at http://www.epa.gov/climatechange/impacts-adaptation/. Accessed July 16, 2012.

Haddow, K. S. (2009). The Case for Adaptation. In G. Bullock, G. D. Haddow, and K. S. Haddow (Eds.), *Global Warming, Natural Disasters, and Emergency Management* (pp. 1–15). New York: CRC Press.

Holland, G. (1997). The Maximum Potential Intensity of Tropical Cyclones. *Journal of Atmospheric Sciences, 54*(21), 2519–2541.

Houghton, J. (2004). *Global Warming: The Complete Briefing.* Cambridge: Cambridge University Press.

Hyslop, A. (2011). The Ontario Emergency Management Act and Municipal Climate Change Strategies: Determining the Relationship. City of Hamilton: Hamilton, Ontario, Canada. Available at http://www.hamilton.ca/NR/rdonlyres/57BB2821-B60C-4B99-8DDF-8307F74BCB41/0/EmergencyManagementAnd-MunicipalCCStrategy.pdf. Accessed June 20, 2012.

Jansen, E., & Contribution of Working Group I of the Fourth Assessment Report of the Intergovernmental Panel on Climate Change. (2007). Paleoclimate. In S. Solomon, D. Qin, M. Manning, Z. Chen, M. Marquis, K. B. Averyt, M. Tignor, and H. I. Miller, (Eds.), *Climate Change 2007: The Physical Science Basis.* Cambridge and New York: Cambridge University Press.

Karl, T. R., Melillo, J. M., & Peterson, T. C. (Eds.). (2009). *Global Climate Change Impacts in the United States.* U.S. Global Research Program. Washington, DC: Cambridge University Press.

Knutson, T., & Tuleya, R. (2004). Impact of CO_2-Induced Warming on Simulated Hurricane Intensity and Precipitation: Sensitivity to the Choice of Climate Model and Parameterization. *Journal of Climate, 17*(18), 3477–3495.

Labadie, J. R. (2011). Emergency Managers Confront Climate Change. *Sustainability, 2011* (3), 1250–1264.

LePage, M. (2007). Climate Myths: Many Leading Scientists Question Climate Change. Available at http://www.newscientist.com/article/dn11654-climate-myths-many-leading-scientists-question-climate-change.html. Accessed February 17, 2009.

Linden, E. (2006). *The Winds of Change.* New York: Simon and Schuster.

Lynas, M. (2007, April 22). Six Steps to Hell. *The Guardian.* Available at: http://www.guardian.co.uk/books/2007/apr/23/scienceandnature.climatechange. Accessed July 16, 2012.

Meehl, G. A., & Contribution of Working Group I of the Fourth Assessment Report of the Intergovernmental Panel on Climate Change. (2007). Paleoclimate. In S. Solomon, D. Qin, M. Manning, Z. Chen, M. Marquis, K. B. Averyt, M. Tignor, and H. I. Miller, (Eds.), *Climate Change 2007: The Physical Science Basis.* Cambridge and New York: Cambridge University Press.

Mileti, D. (1999). *Disasters by Design.* Washington, DC: Joseph Henry Press.

Mooney, C. (2007). *Storm World.* New York: Harcourt Brace.

National Research Council. (2002). *Abrupt Climate Change: Inevitable Surprises.* Washington, DC: The National Academic Press.

O'Brien, G., O'Keefe, P., Rose, J., & Wisner, B. (2006). Climate Change and Disaster Management. *Disasters, 30,* 64–80.

Oreskes, N., & Conway, E. M. (2010). *Merchants of Doubt.* New York: Bloomsbury Press.

PEW Environmental Group. (2008). International Policy Initiatives to Address Global Warming. Available at http://www.pewglobalwarming.org. Accessed June 22, 2009.

Riehl, H. (1950). A Model for Hurricane Formation. *Journal of Applied Physics, 21,* 917–925.

Schneider, R. O. (2006). Hazard Mitigation: A Priority for Sustainable Communities. In D. Patton and D. Johnson, (Eds.), *Disaster Resilience: An Integrated Approach.* Springfield, IL: Charles C Thomas.

Schneider, R. O. (2011). Climate Change: An Emergency Management Perspective. *International Journal of Disaster Prevention and Management, 20*(1), 53–62.

Silverman, J., Levy, L. A., Myrus, E., Koch, K., & DeGroot, A. (2010). Why the Emergency Management Community Should Be Concerned About Climate Change, *CNA Analysis Solutions.* Available at http://www.cna.org/research/2010/why-emergency-management-community-should-be. Accessed July 16, 2012.

Thompson, M., & Gaviria, L. (2004). *Weathering the Storm: Lessons in Risk Reduction from Cuba.* Boston: Oxfam America.

Tomkins, E. C., & Hurlston, L. A. (2005, March). Natural Hazards and Climate Change: What Knowledge Is Transferrable? Tyndall Working Paper No. 69, Tyndall Centre for Climate Change Research, University of East Anglia, Norwich.

Union of Concerned Scientists. (2009). Scientific Consensus on Global Warming. Available at http://www.ucusa.org/ssi/climate-change-consensus-on.html. Accessed May 21, 2009.

Wigley, T. M. L., Ramaswamy, V., Christy, J. R., Lanzante, J. R., Mears, C. A., Santer, B. D., & Folland, C. K. (2006). Executive Summary. In T. R. Karl, S. J. Hassol, C. D. Miller, and W. L. Murray (Eds.), *Temperature Trends in the Lower Atmosphere: Steps for Understanding and Reconciling Differences* (pp. 1–15). Washington, DC: Synthesis and Assessment Product 1.1 U.S. Climate Change Science Program.

Chapter 5

HUMAN-MADE HAZARDS
AND SUSTAINABILITY

We have met the enemy and he is us.

– Pogo

INTRODUCTION

On March 11, 2011, a deadly earthquake and resulting tsunami killed more than 15,000 people in northeastern Japan. It also led to a nuclear crisis at the Fukushima Daiichi power plant, which spewed radiation and displaced tens of thousands of residents from the surrounding area. The Fukushima Daiichi disaster was the worst nuclear accident since the 1986 Chernobyl disaster in the Ukraine. It would be easy to attribute the Fukushima nuclear disaster to an act of nature (i.e., the result of the earthquake and tsunami). But it would not be entirely accurate to do so.

According to a 2012 report issued by the Fukushima Nuclear Accident Independent Investigation Commission, the nuclear disaster was a human-made disaster. It was, according to this report, a result of errors and willful negligence at the plant before the earthquake and tsunami (not wholly unexpected events in this part of the world) struck. It was the result of a flawed response in the weeks that followed the earthquake and tsunami. It was the result of pre- and post-disaster collusion between the facility's operator, regulators, and the government (Wakatsuki and Mullen, 2012). According to the commission, the operator, regulators, and government failed to implement the most basic

120

safety requirements, anticipate such a disaster, assess the probability of damage or prepare for such a disaster, and respond effectively. The causes of the accident were foreseeable prior to March 11, 2011, and many of the tragic consequences of that tragic day were preventable according to the commission (Wakatsuki and Mullen, 2012).

In addition to the specific failings of operator, regulators, and government, the commission also attributed some of the responsibility for the nuclear disaster to Japanese culture and its unquestioning acceptance of nuclear power as an accepted and necessary thing. This unquestioning acceptance was imbedded in the culture to such a degree in fact that it became basically immune to careful and objective scrutiny. The government bureaucracy responsible for its promotion was also its regulator, a situation ripe for ineffectiveness with respect to anticipating, preventing, or limiting the risks of damages such as those at Fukushima Daiichi (Wakatsuki and Mullen, 2012).

The Fukushima nuclear accident is, unfortunately, an example of a human-made disaster that is all too familiar. Historically, it is accurate to say that both corporate leaders and public policymakers everywhere are every bit as prone to dramatic failure as were the operators, regulators, and the government of Japan in the Fukushima incident. They are frequently too quick to take the position of minimizing risks attached to promising new technologies and practices. They tend to overlook or underestimate or even to be willfully negligent with respect to the identification of long-term risks. They are often inclined to neglect investments in safety as they pursue more immediate economic benefits and political objectives (Kirby, 1990). But it is an inevitability that many of the things we might do that constitute a benefit (e.g., the eradication of insects or pests) introduce risks or potential harms (e.g., the poisoning of crops or people) that must be anticipated and managed with forethought (Gallant, 2008). This is to say that sustainable and hazard-resilient communities must of necessity identify hazard risks associated with industrial, technological, and workplace activities and reduce the threats, vulnerabilities, and disaster potential they introduce on the populations they serve.

The practicing emergency manager must anticipate, prepare for, be ready to respond to, and reduce vulnerabilities not only in relation to natural hazards and disasters but also with respect to industrial and technological hazards and disasters. The emergency management profession, to whatever extent it may come to have a unified voice, must promote policies and practices that achieve these ends in the name of sustainability and hazard resilience. As challenging as it is to promote and implement sustainable practices with respect to natural hazards, it is even more difficult to do so with respect to industrial and technological hazards. These (industrial and technological) hazards are not natural. They are entirely human-made, and they are often a risk (or a bunch of risks) that we are willing to take for the sake of progress or for profit. This (i.e., taking risks) is not always a bad thing, but it is always an important thing that must be managed in both the practical work to create and maintain sustainable hazard-resilient communities and the policy arena that supports or frustrates that goal.

The purpose of this chapter is to discuss the special challenges that the industrial and technological hazards we create impose on us and the role of emergency management, the practitioner and the profession, with respect to the identification and management of risks and vulnerabilities in relation to them. The challenges are not unmanageable; they never are really. But there is an added degree of difficulty given the economic and political stakes that are attached to the hazards we create. This often means that the work of hazard mitigation, disaster preparedness, and even disaster response is made more difficult and contentious than is healthy for sustainable hazard-resilient communities.

We will discuss in some greater detail the phenomenon we have labeled hazard creation. From there, we shall proceed to discuss two specific cases that exemplify both the challenges such hazard creation may present and the difficulty of managing them. We will conclude with a discussion of the role that emergency management, as a sustainability profession, must play in relation to these hazards of our own making.

CREATING HAZARDS

Industrial and economic development, good things in general to be sure, often require the taking of risks. As technology permits us to act and create on a larger scale, it may as an unintended byproduct introduce new heretofore unanticipated challenges. But that too is something we can, if we choose to do it, manage successfully. Consider the vast array of hazards created by our technologies and our industrial advances over the decades that are now commonplace and considered manageable by us, so much so that we scarcely give them a thought.

Many industrial and workplace hazards are well known and do not pose unanticipated or unmanageable consequences (i.e., we have learned to anticipate and control for them). These common hazards include toxic substances, flammable materials, explosive materials, excessive noise, corrosive materials, biologically active materials, heat or cold stress, oxygen deficiencies, accidents resulting in physical harm, radioactive materials, and cancer-causing agents. The common consequences of toxic substance exposure may include, for example, asphyxiation; poisoning; cancer; damage to liver, kidneys, and nerve cells; harmful effects to unborn children; loss of limbs; skin diseases; loss hearing; and eye injuries (Strong and Irwin, 1996). That these are, in effect, everyday hazard potentials introduced by our technological and industrial advances and that we choose to live with the risks they impose, are willing to accept them, and are able to manage them safely is perhaps a tribute to our skill at anticipating and managing the negative byproducts of our technology. Concise integration of scientific, environmental, and management research has led to the development of successful mitigation efforts, emergency response plans, and regulatory strategies that protect public and worker safety. But two things are easily forgotten as we compliment ourselves on our ability to manage the risks associated with technological and industrial progress. First, it often took either a catastrophe (e.g., an industrial disaster of an unprecedented sort) or a chronic condition created by cumulative impacts to make these risks known to us. Second, it always took a concerted effort to overcome the resistance of the hazard creators (industries, corporations, politicians, etc.) to public efforts to manage these hazards with appropriate regulatory policies and structures.

In fairness, sometimes the creation of hazards may be unknown to some extent, and the resulting disasters may be genuinely unexpected. Such a situation may require a precursor or first-of-a-kind event to make the hazard fully known. The first commercial jet airliner crash, for example, the British Comet, may have been necessary to discover the full potential of the metal fatigue problem (Mitchell, 1996). But it would be a mistake to take the position that hazardousness in general is typically a surprise or that precursor first-of-a-kind or superlative worst-of-a-kind events are always necessary to predict and control risk. In other words, it is intellectually lazy to attribute industrial disasters to unexpected malfunctions, unknowable risks, or unanticipated side effects of technological systems. "The calculus of industrial hazards is a blend of industrial systems, people, and environments" (Mitchell, 1996, p. 12). Hazard risk is most often knowable and predictable. The combination or interaction of industrial systems (facilities, equipment, products, waste), people (operators, managers, populations), and environments (atmosphere, hydrosphere, lithosphere, biosphere) produces a predictable disaster event.

Even where great effort is taken to anticipate hazards and prevent disasters, where safety is a shared concern, human decision making is imperfect and may contribute to a catastrophic outcome. The tragic Challenger Space Shuttle disaster, for example, was ultimately blamed on a vulnerable fluid seal. That makes it seem like the tragic explosion during the launch was an unforeseen accident, but that perception is not totally accurate. In fact, it was well known before the "accident" that the fluid seal was compromised in cold weather and subject to failure. It was further known that the failure of this seal could be fatal. On launch day, a cold day – cold enough in fact to cause concern to be expressed about the seal – a human decision was made to launch despite the known hazard and the risk it portended. To its immense credit, NASA is an agency known for its dedication to safety. But it was, in 1986, also under growing political pressure to keep its launch pace on schedule. It was sensitive to criticism it had received for what were considered by its critics to be an excessive number of delays and cancellations. Delays and cancellations were frowned on and to be avoided if at all possible. A decision was made to launch. The resulting disaster was a combination of factors (i.e., a vulnerable fluid seal, cold weather, an impatient launch team), but it was the human element (the decision to take the risk in a context known to be hazardous) that created the disaster (Vaughan, 1996). This is generally the case where the hazard and the risks associated with it are known in advance, which is in fact most of the time.

Practicing emergency managers have long known that concerns about industrial and technological disasters are part of their everyday work experiences. A prime example of the mainstreaming of this concern into their work is the Superfund Amendments and Reauthorization Act (also known as SARA Title III) of 1986 and the Emergency Planning and Community Right to Know Act (EPCRA) contained therein. This statute was designed to improve community access to information about chemical hazards and facilitate the development of chemical emergency response plans by state, tribal, and local governments. EPCRA required the formation of State Emergency Response Commissions (SERCs). The SERCs were responsible for coordinating emergency responses in relation to chemical disasters. They also were responsible for appointing Local Emergency Planning Committees (LEPCs). State and local governments were to assess the risks of chemical hazards (e.g., transportation, storage, releases, etc.), take actions to prevent disasters, and prepare plans for response to any disasters that might reasonably be anticipated. EPCRA required facilities (industries, businesses, farms, etc.) to notify the SERC and the LEPC of the presence of any hazardous substance (a list of specified substances was provided) in excess of a specified threshold. They were required to submit inventories of toxic chemicals to local authorities to facilitate emergency response planning, submit toxic chemical release forms that reported all transfers of chemicals, notify the SERC and the LEPC in the event of a release (accidental or other) of toxic chemicals, and appoint someone to coordinate with the local LEPC.

In the process of complying with the Emergency Planning and Community Right to Know Act, many local communities created hazardous materials response teams (HAZMATs) and capacities. Almost every practicing emergency manager can tell you that, given the materials transported through the railways and roadways in their community, they are one train derailment or one tanker truck accident away from a catastrophe. They can also tell you about the various hazard potentials represented by industries and businesses that utilize and store chemicals in their community and of the preparedness plans they have in place to respond to any hazardous materials disaster. This, as much as any natural hazard or disaster, is part of their work. They all have some experience coping with and planning for the human-made hazards that are a by-product of industrial and technological advancement. But if they are honest, they can also tell you that they are not quite comfortable or certain that they are able to keep pace with the creation of these hazards or to be confident that some of the people who make decisions about which risks to take or avoid are motivated by public and worker safety or disaster prevention and hazard resilience as a first priority.

To some extent, emergency managers like the rest of us simply deal with industrial disaster outcomes that are the product of actions and decisions about hazards and risks over which they have no control and little influence. Those who do make the decisions (in the private and public sectors alike) are inclined to minimize the hazards and risks they may impose. This does not mean that they choose to be reckless in every single case, but it does mean that both economic and political factors often narrow their focus to immediate short-term benefits to the exclusion of an analysis of longer term threats and risks. The threats that this may impose on communities with respect to sustainability and resilience are significant enough to warrant the urgent attention of all who care about hazard resilience and the emergency management profession. Two specific examples make excellent case studies in this regard, and we turn our attention to them now for purposes of elaboration.

The first case involves an industrial setting where the hazards created, the risks taken, predisaster preparedness, and most of the disaster response activities are primarily the responsibility of the industry itself. Emergency managers would have a minimal support role to play in a disaster scenario. But the disaster impacts the health of the community as well as its economic vitality and environmental sustainability. As such, it should be of more than a passing interest to an emergency management profession, which presumably has insight and experience relevant to the private sector as well as the public. The second case involves a setting in which the hazards created and the risks taken are the product of a relatively unregulated industry, but the responsibility for predisaster planning and disaster response will in fact involve the practicing emergency manager more directly. As with the first case, any resulting disaster impacts would adversely affect the health and safety of residents of the community as well as its economic vitality and environmental sustainability.

CASE ONE: THE DEEPWATER HORIZON DISASTER

The extraction of oil from off-shore wells in the Gulf of Mexico has been common since the end of World War II. Most of the wells were initially in shallow water, about 200 feet deep. In the event of a problem or an accident, divers could be sent down to fix them. But these inshore wells became depleted in time, and the oil companies went into deeper waters to extract oil. They pushed farther and deeper off shore (more than 5,000 feet beneath the surface in the case of the Deepwater Horizon), and, as they did so, they escalated risks and increased safety concerns considerably. Divers cannot fix deep wells. Robotic instruments must be used in their place at deeper levels and with considerably less precision than human divers in an environment (i.e., different chemistry of the water) where the fixes that work in shallow water do not work (Southern Studies, 2010). It is generally agreed that as oil companies have gone to greater extremes to acquire oil from off-shore wells, they have created new hazard potentials, increased considerably the level of risk, and underinvested in appropriate safety precautions.

Congressional hearings in July 2010 concluded that the oil industry had spent billions of dollars to research and develop the technologies and practices for deep water drilling, but that little investment had been devoted to the technologies for accident prevention and hazard mitigation. Despite well-known and increased risks, and the increased prospect for high-impact accidents or disasters, oil companies were in general not willing to make the necessary investments in safety because these might cut into their profits (Energy and Environment Subcommittee, 2010). A lax attitude toward safety and a culture of maximizing profits at its expense is not, unfortunately, the exception in the oil industry.

British Petroleum (BP), owner of the Deepwater Horizon site, had a history spanning decades of reckless, willfully negligent, and illegal behavior with respect to safety. Prior to 2010 and the Deepwater Horizon explosion, its best-known disaster had been the 2005 explosion of a refinery in Texas. This refinery explosion in Texas City (near Galveston) killed 15 workers, injured 180, and endangered thousands of nearby residents. A subsequent investigation by the U.S. Chemical Safety and Hazard Investigation Board found organizational safety deficiencies at all levels of the corporation. BP pled guilty to a felony violation of the Clean Air Act and was fined $50 million (McClatchy, 2010).

In Alaska, more than twenty-five years ago, chronic BP safety deficiencies came to light in the aftermath of the Exxon Valdez oil tanker spill. Exxon and BP were partners in Alaska's Prudhoe Bay oilfield and shared ownership of the trans-Alaska pipeline system (Alyeska). Alyeska routinely failed to live up to BP's promises to limit and contain accidental spills. Its North Slope Corrosion Control Program, for example, failed miserably time and time again. Despite warnings from a leak detection system, a corroded piece of pipeline in Prudhoe Bay was ignored as it leaked more than 200,000 gallons of oil over a five-day period in March 2006. A smaller leak occurred five months later. Investigations by Congress, prompted by these events, found that the entire line was riddled with corrosion, and BP workers were actively discouraged from reporting this or any related safety and environmental concerns. Many other disturbing reports were indicative of a lax attitude toward safety and the environment (McClatchy, 2010).

After the 2005 Texas City refinery explosion, the U.S. Chemical Safety and Hazard Investigation Board issued a report that documented BP's history of ignoring warning signs, willfully failing to invest in safety as a means of cost-cutting, and pushing their employees to maximize profitability at the expense of both worker and public safety (McClatchy, 2010). There was some evidence of an increase in safety investments after the 2005 Texas City explosion, but a corporate culture of cost cutting and cutting corners on safety to boost profits was found to have persisted (Hanson, 2010). Given its record, one would be hard put to say that BP was doing an adequate job of identifying the risks and vulnerabilities associated with its work and managing them effectively in the interest of public and worker safety.

The BP Deepwater Horizon disaster was not a surprise or an unexpected event. Remember, as we have already noted, "the calculus of industrial hazards is a blend of industrial systems, people, and environments" (Mitchell, 1996). To call Deepwater Horizon an accident or to blame it on faulty equipment is to ignore the element of human judgment, very poor judgment at that, and to ignore everything that could have prevented the disaster or reduced its likelihood. What has come to light and is now known about the events of April 20, 2010, just prior to the explosion and what we have come to know in the aftermath of these events tell a very human story about another disaster by human design.

On April 20, 2010, an explosion at the Deepwater Horizon drilling platform killed eleven people, unleashed an environmental catastrophe in the Gulf region, and adversely impacted the Gulf economy. Initial investigations quickly focused on a piece of equipment that had failed, the blowout preventer (BOP). The BOP is a large mechanism consisting of a series of high-pressure hydraulic valves designed to control the flow of oil and gas from the well. Any uncontrolled flow may cause a well to blow given the extreme pressures at the ocean depths in deep water drilling sites. As the pressure built and the BOP was engaged to control the flow, the blind-sheer ram (which uses two blades to cut through the metal pipe and seal the wellbore) failed to engage. It is worth noting that the reliability of blind-sheer rams had been repeatedly questioned and deemed insufficient by a number of studies conducted over the decade preceding Deepwater Horizon (Energy and Environment Subcommittee, 2010). It is also worth noting that failed equipment is not the entire story of this disaster.

In the hours before the explosion, the crew at Deepwater Horizon noted various and serious warning signs that should have alerted them to the potential for an explosion. The well was about to be sealed as per normal procedure by Transoceanic, the company contracted to do the drilling. BP was eager for Transoceanic to complete its work as the drilling operation cost BP almost $1 million per day. The normal procedure at the completion of the drilling would be to remove the heavy drilling lubricants and replace them with lighter fluid before sealing the well off until BP was ready to extract the oil.

According to the House Energy and Commerce Committee report issued a month after the explosion, crews noticed unusual and potentially dangerous pressure and fluid readings that caused them considerable concern (Energy and Commerce Committee, 2010). Removing the heavier drilling fluids would have been counterindicated by these readings. In addition to troublesome pressure and fluid readings, there were concerns about the BOP. Evidence had suggested that it may have been damaged in an earlier incident, and it was not certain to crew members and the crew chief that it would do its intended job. BP executives and drill hands apparently debated the wisdom of proceeding as planned with the removal of the heavier drilling fluids. Five hours before the explosion, an unexpected loss of fluid was observed, thus suggesting that there were leaks in the BOP. Over the objections of the rig's chief mechanic, a BP executive ordered the removal of the heavy fluid and its replacement with lighter weight sea water as per normal procedure. In the face of what was known on the night of April 20 and despite obvious and unmistakable warnings that something was amiss, decisions were made to proceed as per normal. One would suspect that the basic decision was that it was worth rolling the dice on safety to cut costs (Bolstad, Goodman, and Taylor, 2010).

One would expect that, given the risks associated with drilling for oil thousands of feet beneath the ocean, federal regulators would take extra care to ensure that BP, a company with a corporate culture and a history of cost cutting where safety was concerned, would take all of the necessary precautions to assess and reduce risks and make all necessary preparedness efforts to respond to a possible disaster. But such an expectation would be disappointed as oil companies generally create their own off-shore safety rules, are generally lax with respect to disaster preparedness plans, and, due to various protections of proprietary interests, are not required to share the technologies they have developed for disaster response. This is more or less the case internationally. The United States is not alone in ceding responsibility to the oil industry for creating their own rules and for the design and implementation of safety features in off-shore rigs. Beyond a general lack of safety regulation and the reliance on oil producers to self-regulate and invent appropriate safety technologies, the U.S. governmental agency with oversight responsibilities for offshore drilling failed miserably in meeting even minimal expectations of professional and public responsibility.

The Minerals Management Service (MMS) was the bureau within the Department of Interior charged with the responsibility for monitoring the development and extraction of mineral resources in the federal waters off the U.S. shores. They were responsible for the oversight and inspection of drilling sites. Admittedly, the regulation of drilling sites may have been minimal, but they were expected to meet some specified safety standards. These were to be enforced through safety inspections by the MMS. But the MMS had only sixty inspectors to look after 3,800 platforms in the Gulf. Many of these inspectors were not properly trained, according to congressional investigators, and inspections were haphazard at best (Thomas, 2010). It should also be noted that chronic safety problems were the norm on all of the platforms throughout the Gulf long before the Deepwater Horizon disaster. From 2006 to 2009, thirty platform employees working in the Gulf of Mexico were killed in accidents, and 1,300 people were injured. Workers died in fires, fell through holes in platforms, and were crushed and killed by falling pipes. Despite chronic safety problems, the MMS imposed only a few paltry fines that often took years to collect. In the overwhelming majority of cases where workers were actually killed, there was no record of fines being paid (Thomas, 2010).

As if minimal regulations sporadically and ineptly enforced were not enough, the MMS was found to have a culture of generally lax oversight and a history of excessively close ties with the oil industry. Documented reports of misdeeds and negligence included the acceptance by MMS officials of gifts from the industry, sexual relations between industry officials and MMS staff in the leasing and inspection office in the Gulf region, the falsification of inspection documents, and a history of lax and inadequate performance generally (Urbina, 2010). In the aftermath of the Deepwater Horizon disaster, these documented reports led to the disbanding of the MMS and the division of its work among three other bureaus.

The trail of flawed human decision making (industrial and governmental) that contributed to the Deepwater Horizon explosion in the Gulf is compounded by human decisions made during the response phase of the disaster. An examination of the immediate response to any oil spill or blowout disaster contains its own horror stories that only make a bad situation worse. What generally ensues in the immediate aftermath of such a disaster amounts to a sort of response theatre. It is defined by efforts to contain the damage to the company image, contain any negative political fallout, and limit legal liabilities to whatever extent is possible. Oil companies, which are the main first responders to the disaster, are more inclined to follow a script rather than deal forthrightly with the public or government. The main features of that script are as predictable as the disaster, and it contains most of the following elements:

- understate the amount of oil spilled or released;
- control access and prevent or delay independent analysis;
- understate the environmental impact of the disaster;
- overstate the effectiveness of the company's response effort;
- deny or obscure any long- or short-term damage to the community;
- buy off the locals (i.e., offer money to locals impacted in exchange for waivers promising not to sue for damages);
- slap gag orders on workers or volunteers (cleanup workers especially are to be kept away from the media);
- understate or deny any health risks posed to cleanup workers or community members by things such as chemical dispersants or exposure to the oil itself;
- spread the blame to others (e.g., partners, nature, God); and
- spend millions of dollars in a massive PR campaign to reassure the public and to control damage to the corporate image.

The response theatre script works, not perfectly perhaps but well enough to accomplish several industry goals. None of these goals is necessarily related to efficient disaster response. The first goal of the oil company is to protect itself against any charges of negligence or misconduct.

As an early part of its strategy, BP sought to evade or at least limit its responsibility for the Deepwater Horizon explosion, spread the blame to Transoceanic (the company that owned and operated the drilling equipment) and Halliburton (manufacturer of the complicated safety equipment that had failed), and shift the focus away from any miscalculations or mistakes on its part (Hanson, 2010). In addition, it appears that BP intentionally underestimated and/or underreported the amount of oil rushing into the Gulf. On April 24, BP announced that only 1,000 barrels per day were leaking into the Gulf. On April 28, BP raised the estimate to 5,000 barrels per day. The estimate continued to creep upward to 12,000 barrels per day, to 19,000 barrels per day, to 35,000, and, by mid-June, to 60,000 barrels per day (Hartenstein, 2010). The estimates began to increase only after the public was allowed to have unrestricted viewing of the underwater camera BP had trained on the site. Initially, BP had allowed the viewing of only short periods of video from limited angles, and this made independent analysis of the flow difficult. When Congress finally mandated unrestricted viewing, BP's estimates increased as independent experts began to disprove the earlier lowball estimates (Hartenstein, 2010). The changing estimates suggest that BP's original estimates may have been part of an intentional effort to control information and minimize public awareness of the seriousness of the damage.

In both the immediate and long-term response to a disaster, the oil company's major focus will be on protecting its public image and reducing legal liabilities. Its number one concern is the disaster's impact on management, shareholders, and employees. This may be understandable, perhaps even quite logical. Company lawyers would be negligent if they did not advise their clients to take all steps necessary to avoid legal liabilities. It is also logical that a massive PR campaign will be rolled out to promote the company's image and create an impression that everything is under control, the disaster is minor, the company cares, and the future is bright. But it is interesting to note that as BP provided such assurances, it did not initially have a clue how to stop the leak. BP had not prepared for a disaster of this magnitude, hence it took a great deal of time and trial and error before a solution was found. It had not adequately assessed and prepared to manage the risks associated with the hazards their deep water drilling technology created. It made a series of decisions to take risks with safety, and its response activity was as much (more in fact) a theatrical production to protect its image than it was a well-planned and efficiently implemented effort to deal effectively and honestly with a disaster of its own making.

The conclusion of the federal investigation into the Deepwater Horizon disaster emphasized that BP and its contracting companies (Halliburton and Transoceanic) were jointly responsible for the disaster. Among the contributing causes specified in the federal investigation were BP's cost-cutting and time-saving decisions that failed to consider risks, contingencies, and mitigation steps; failure of the rig crew to stop work after encountering multiple hazards and warnings; BP's failure to adequately assess the risks associated with the operational decisions it made leading up to the blowout; BP's failure to ensure that all risks associated with its operation at the Deepwater Horizon site were as low as possible; BP's failure to adequately supervise and have accountability for all activities associated with the Deepwater horizon; BP and Halliburton's failure to perform the production casing cement job adequately and in accordance with industry-accepted specifications; and BP and Transoceanic's collective misinterpretation of negative equipment tests (Johnson, 2011). The final report noted that, in addition to the specific instances of poor risk management, generally insufficient response training handicapped those making critical decisions on the night of April 20, 2010 (Johnson, 2011).

In both the pre- and postdisaster phases of the Deepwater Horizon disaster, what was entirely missing was what one might call an emergency management perspective. Appropriate applications of knowledge to hazard identification, risk reduction and mitigation, emergency planning, response preparedness, and safety were in short supply. Emergency management in an industrial setting is frequently frustrated by a corporate culture and a recurring cycle of decisions that work against the application of knowledge and expertise to the tasks of risk reduction and safety as a first priority. Management incentives and rewards are not centered on safety and risk reduction, and this is all too commonly reflected in the cutting of corners and costs before a disaster occurrence, the lack of preparedness for a disaster occurrence, and the lack of transparency and openness during and after a disaster event.

Among the least noticed but most important aspects of a disaster such as the Deepwater Horizon, at least as far as the general public is concerned, is the long-term environmental and human damages that may result. For example, it has been nearly a quarter of a century since the Exxon Valdez tanker spill, but we are still learning more about its ongoing and ultimate impact. In March 1989, when the Valdez ruptured her hull on Bligh Reef and spilled some 30 million gallons of crude into Prince William Sound, she left a devastation in her wake that is still being totaled. In addition to the thousands of marine animals and millions of salmon and herring that were among her immediate victims, the years to follow have seen continued and ongoing threats to plant and animal life. They have also revealed a trail of pervasive health problems for cleanup workers related to the use of chemical dispersants in the cleanup phase. The impact of the Exxon Valdez continues to unfold many years after it has become a long ago and forgotten incident (Ott, 2005). It will be many years before the full impact of the Deepwater Horizon disaster is known as well. According to the initial analysis, it caused measurable and extensive short-term damage to marine and wildlife habitats and to the Gulf's fishing and tourism industries, but the long-term damage is not yet known.

Given the immense potential a disaster such as Deepwater Horizon may have to affect the environment, health, and sustainability of a community over an extended time period, it is inconceivable that it would not be a matter for study and discussion within an emergency management profession. But before addressing this thought, and the contribution that an emergency management profession might make to the discussion, let us look at another case, one that has more immediate implications for practicing emergency managers with respect to the introduction of new hazards and risks with which they may have to deal more directly.

CASE TWO: THE FRACKING REVOLUTION

Hydraulic fracturing (hydrofracking or fracking) is the fracturing of rocks far beneath the earth's surface for the recovery of natural gas. This technique for natural gas extraction has been in use for more than fifty years. But in the early twenty-first century, technological innovations have made horizontal drilling techniques feasible, and these new techniques have greatly expanded the potential for natural gas extraction as well as its profitability. This new drilling technique has also introduced (predictably of course) new hazard potentials to be considered and managed.

Fracking has generated both staunch support and significant criticism. Supporters have applauded the new technologies that have opened up the potential for the accelerated exploration for natural gas. They have praised natural gas as a cheap, clean, abundant fuel for the future. Some critics have suggested that the new technologies associated with the horizontal drilling for natural gas will have large and undesirable environmental effects and pose significant risks to public health (Botkin, 2010). New natural gas discoveries suggest that the United States is awash with reserves that can be used to substitute for coal in power plants and serve as a bridge to a low carbon future and as a transition fuel in the ongoing battle against climate change. But concerns about the possible risks associated with hydraulic fracturing (i.e., fracking) have escalated as this method of natural gas extraction has become more commonplace and its effects have been debated.

Fracking presents a number of hazard potentials and likewise holds the potential to increase a number of risks that practicing emergency managers are already expected to prepare for, adapt to, and cope with in their work. This is especially true, as we shall see later, with respect to hazards and risks associated with chemical use, storage, transportation, and toxic wastes. Practicing emergency managers will want to see themselves as significant stakeholders in the communities where fracking is taking place. Likewise, given the broad debate across governmental levels (national, state, and local) about the risks associated with fracking, informed input from an emergency management perspective would be of practical value (a necessity in fact) for policymakers. Hence, a role for emergency management as a profession is also clearly implied. The research on the risks associated with hydraulic fracturing is in its infancy, relatively speaking, but it is beginning to raise some important concerns that natural gas producers and politicians have been slow to engage. The concerns raised are pretty basic and straightforward to emergency management professionals, public health officials, and scientific researchers. Before addressing these concerns and the need for responsible analysis and timely policy in relation to them, let's do a quick overview of the fracking process.

At present, natural gas provides about 22% of the energy in the United States and about 26 percent worldwide (Botkin, 2010). In addition to conventional sources of natural gas, unconventional sources (i.e., ocean deposits of methane hydrates, coal-bed methane, and shale gas deposits) may be able to provide larger amounts of fuels than previously thought. These unconventional sources, targets made ever more exploitable by the new horizontal drilling technology, are understandably inviting to energy producers. In July 2009, the U.S. Department of Energy announced that estimated U.S. natural gas reserves were 35% larger than previously estimated (Botkin, 2010). Much of this abundance can be attributed to the new possibilities for tapping into unconventional sources. Together these unconventional sources may provide as much as 60% of all natural gas in the United States by 2035. This suggests that they will be prime targets for accelerated application of new technology. The most inviting target presently would appear to be shale gas deposits.

Shale is one of the most common kinds of rock in the United States (found in twenty-three states) and, with recent technological developments, holds particularly great potential for natural gas exploration. Shale gas deposits presently provide about 25% of the natural gas in the United States but, with both the abundant supply and new hydraulic fracturing technology, are expected to provide 45% by 2035 (Katusa, 2011). This trend will be accentuated by the unquestioning popularity of natural gas as a supposedly clean and affordable energy alternative. The abundance of supply from unconventional sources has thus spurred a rapid development of the large deposits of shale gasses in the United States. It is accurate to say that what has resulted might be called a revolution in natural gas exploration (Maugeri, 2010).

In the first decade of the twenty-first century, shale gas production has literally exploded. There has been a rapid, and relatively unregulated, expansion of shale gas production in the United States. It began in Texas (the Barnett Shale Field) in 2000 and has led to a race to leverage immense shale deposits around the country. The two best-known deposits may be the Hainesville Shale in Louisiana and the Marcellus Shale that stretches from West Virginia through Pennsylvania and New York. U.S. shale gas production jumped from almost zero to about two trillion cubic feet between 2000 and 2008 (Maugeri, 2010). As the boom in shale gas exploration took off, the conversation about the new horizontal fracturing (i.e., fracking) technology and any potential risks it may pose to public health and safety has lagged behind as natural gas producers, public policymakers, and media have all been and remain relatively silent on the topic. The boom took off, and it has enjoyed unquestioned support from policymakers and the public, been largely accepted as a good thing, its hazard potentials and risks never fully explored, and it has never been subjected to significant safety regulations.

Hydraulic fracturing or fracking is a drilling technique that involves pumping large volumes of pressurized water, sand, and chemicals into deep shale deposits. This pumping is done to fracture the rocks and release the oil or gas. While some drillers have been fracking since the 1950s, the last decade has seen advancements in technology that have taken this drilling technique to new levels (Katusa, 2011). The major technological advancement has been related to new horizontal drilling techniques that have enabled producers to extract gas from deposits that used to be inaccessible. Fracking had been used in vertical wells for some time, but it did not retrieve enough shale gas at economic levels. But as drilling advanced to where drillers were able to frack horizontally, it broadened greatly the potential for extraction from a single well and improved its profitability in no small measure (Katusa, 2011). While popular and profitable, it is important to know that this new technique of natural gas exploration is not without some potential and significant risks to human and animal populations.

Millions of gallons of water mixed with sand and fracturing fluids (i.e., chemicals) are required to frack a well. A well may be fracked up to eighteen times (Katusa, 2011). As we have already noted, this process fractures the rocks and releases the oil or gas. With each fracking treatment, about half of the fracking liquid returns to the surface with the gas (via collection pipes), and about half remains in the ground. The retrieved gas is piped to compressor stations, purified, and compressed for transport. The returning fracking fluids, now called wastewater, are handled in a variety of ways. They may be transported to water treatment plants (which are not generally designed to handle or treat fracking fluids), they may be stored in large tarp-lined pits and be allowed to evaporate, or they may be re-injected into old wells (Katusa, 2011). As one might expect, the fracking fluids are a primary source of concern: their chemical composition, concerns about their usage and disposal, the possibility of chemical or waste spills during transportation, the potential risk of polluting water tables needed for drinking water and agricultural use, and other potential public health-related impacts.

Many fracking fluids can be toxic to humans and wildlife. These chemicals are known to cause cancer. Chemicals used in fracking include, potentially at least, benzene, toluene, boric acid, xylene, diesel fuel, methanol, formaldehyde, and ammonium bisulfate (Earth Works Action, 2011; Katusa, 2011). It should be noted that drillers are not required to report or make public the formulas for the chemical cocktails they create and use. This is in deference to the industry's proprietary interests associated with the manufacturing of the chemical cocktails utilized in the fracking process. The potential for the contamination of groundwater from these chemicals exists primarily due to the possibility of leaks through cement well casings (over time, these may deteriorate and crack). Most of the fluid remaining in the ground is lower (i.e., 5,000 to 8,000 feet beneath the surface) than groundwater aquifers that are generally no more than 1,000 feet below the surface. But the potential for cracks in cement well casings and the escape of chemicals or methane gas during the process of insertion and extraction is real. This could pose a threat to groundwater aquifers. In addition to the fracking chemicals or fluids, the impact of potential methane gas leaks (potential for explosion and asphyxiation) is an important concern in relation to ground wells in rural areas. Finally, the potential for errors in waste disposal or improper treatment of the retrieved wastewater are among the other major concerns associated with the relatively unregulated acceleration of horizontal fracking (Revesz, Breen, Baldassare, and Burruss, 2010; Zoback, Kitasei, and Copithorne, 2010).

Energy producers are quick to deny that any of the risks associated with fracking represent significant concerns. They reassure us, as a matter of routine but often without the rigorous science to back up their reassurance, that all risks are minimal and manageable. This is to be expected. The energy industry has a long history of developing new technologies that expand risks as it identifies the potential for greater profits. Companies often embrace risk, believing they can either avoid it or safely roll the dice as they seek to expand their market presence and reap the gains to be had by being aggressive. Their success in risk-taking adventures is often aided in circumstances where the negative impacts may take thirty to forty years to be felt. Being aggressive, of course, is not necessarily a bad thing. However, being reckless with respect to risk assessment and risk management is always a bad thing. But risk assessment leads, ideally, to risk management (i.e., mitigation) and perhaps even necessary governmental regulation for public health and safety. From the business and corporate perspectives, safety is costly and, together with regulations, may reduce profits. Energy producers are thus frequently more inclined to cut costs, including safety-related costs, to maximize profits than they are to assess and manage risks.

Like most organized interests and all corporations, energy producers work hard to influence the policy process. They spend great amounts of money to avoid governmental regulation and, in effect, bury risks associated with their work. Corporations and fossil fuel interests have utilized a handful of dissident scientists, for example, to cast doubt on the likelihood of adverse impacts arising from global climate change. These scientists who oppose the scientific consensus about global climate change have been funded primarily by the fossil-fuel industry to create the illusion of uncertainty (Beder, 2002). This support of "experts" who promote interpretations of research outcomes (most are not doing their own research on climate change but are instead criticizing the science) desired and paid for by the industry is intelligently combined with sophisticated political and public relations campaigns designed to reduce the visibility of risks and, thus, the likelihood of governmental regulation. Corporate financing and use of front groups (i.e., corporate-generated grass roots responses) provide a cover of community concern for corporate interests. These front groups also give elected officials the appearance of responding to voters rather than voting for (or being the tool of) corporate interests (Beder, 2002; Merrill, 1991). These practices have been common and ever more effective over the past two decades. Whether restoring the image of an industry (Lindheim, 1989) or promoting its interests in avoiding governmental regulation or weakening public awareness of environmental threats posed by their activity (Beder, 2002), corporations spend immense resources to shape public opinion and influence public policymakers.

Trying to change or influence the way the public and politicians think, a legitimate practice to be sure, may include as a matter of unfortunate routine corporate efforts to weaken public awareness of environmental threats or public health risks. Indeed, corporate responses to scientific research that point out these threats or risks often includes millions of dollars spent for cover-ups, deceptions, data manipulation, fraudulent claims, and fake studies (Beder, 2002). Thus, governments must play a proactive role in monitoring and regulating for public health and safety. But governments, under the influence of corporate lobbying and public relations campaigns, are often reluctant or tardy with respect to meeting this responsibility.

The 2005 Energy Policy Act passed by Congress (crafted by Vice President Cheney, who once ran Halliburton, one of the companies that pioneered fracking) exempted hydraulic fracturing (fracking) from meeting the requirements of the Clean Air Act, the Safe Drinking Water Act, and the Clean Water Act (Katusa, 2011). This Act was preceded by a 2004 determination (that was neither comprehensive nor scientifically rigorous) by the Environmental Protection Agency (EPA), which concluded that the extraction of natural gas via horizontal fracking posed little to no threat to drinking water or public health. This study was denounced by at least one EPA whistleblower for its poor science and as having been the product of an industry-influenced review panel (Bruzelius, 2010). In 2010, the EPA reversed this earlier stance and announced it would launch a $1.9 million research program to assess public health risks associated with fracking (Bruzelius, 2010). But as this study is conducted over the next several years, it will likely not slow production in the least.

From 1960 to the present, there have been more than one million fracking wells in the United States. The natural gas industry claims that this drilling has not caused a single case of groundwater contamination. The industry is adamant that groundwater contamination is not a risk associated with the new horizontal fracking technologies. This is not true according to the Pennsylvania Department of Environmental Protection, which has documented the contamination of an aquifer that fills household wells in a rural area where more than sixty wells were drilled in a nine-square-mile area (Katusa, 2011). There are other such reports from Pennsylvania to Colorado of possible groundwater contamination (Bruzelius, 2010). Recent studies from New York assert that improperly treated fracking wastewater (containing radioactive materials and harmful chemicals) is finding its way into the state's bodies of water (Skrapits, 2011).

An important study concerning the methane contamination of drinking water in conjunction with hydraulic fracturing was published in the spring of 2011 (Osborn, Vengosh, Warner, and Jackson, 2011). The researchers identified the specific fracking concerns related to drinking water (i.e., toxicity of produced water from fracturing fluids that may be discharged into the environment, fluid and gas flow and discharge into shallow aquifers, the impact on private wells that rely on shallow groundwater for drinking and agricultural use, and the potential for explosion) and proceeded to conduct tests in Pennsylvania of drinking water wells in the proximity of fracking activity. Sixty wells were tested, and methane concentrations were found in fifty-one (85%) of them.

The average *methane concentration in shallow groundwater in active drilling areas was seventeen times higher and exceeded the level identified for "urgent hazard mitigation" by the U.S. Office of the Interior.* In this study, there was no evidence of contamination of drinking water by fracking chemicals or fluids. But the correlation of drilling and high methane levels was considered a cause for heightened concern (Osborne et al., 2011).

The authors of the Pennsylvania methane contamination study recommended the long-term monitoring of the industry and private homeowners. They urged drilling firms to comply with a recent request by the EPA to voluntarily report the constituents of fracking fluids. Some have subsequently complied with this EPA request. Most importantly, they called for systematic and independent data collection on groundwater quality before drilling begins and as it proceeds in any region and stressed the need for greater stewardship, more knowledge, and sensible regulation to ensure the sustainable future of shale-gas extraction (Osborne, Vengosh, Warner, and Jackson, 2011).

In an interesting aside to the methane contamination study, it can be noted that some residents in Pennsylvania reported that their water wells had exploded or could be lit on fire. The drilling industry and its supporters described these cases as anecdotal. They also said the findings of the methane contamination study were unconnected to drilling activity. Clearly, and despite this perhaps predictable industry response, it would at a minimum be wise to continue investigating such widespread cases of methane contamination.

Another new study released in the spring of 2011 called into question the notion of natural gas as the cleaner energy alternative. It also cast significant doubts on its benefits in combating global warming (Howarth, Santoro, and Ingraffea, 2011). The authors of this study concluded that the greenhouse-gas footprint of natural gas is actually greater than that for conventional gas and oil or coal. They demonstrated that when you look at the footprint of shale over a longer time span and include the assessment of waste, leaks, production technology, and consumption, a natural gas well will, over the course of its lifetime, contribute more greenhouse gas emissions than previously thought. In fact, the overall carbon footprint of shale gas will be 20% greater than coal according to their analysis (Osborne et al., 2011). Other recent studies have also suggested that methane has greater global warming potential than previously assumed (Jamarillo, Griffin, and Matthews, 2007; Shindell, Faluvegi, Koch, Schmidt, Unger, and Bauer,

2009; Shires, Loughran, Jones, and Hopkins, 2009), thus challenging the notion of natural gas as the cleaner energy alternative or a bridge fuel to a cleaner energy future in the battle against climate change. The natural gas industry has questioned the accuracy of all these studies. In fact, on the release of any new research on the risks or negative environmental impacts associated with fracking, the industry is quick to question the legitimacy of the conclusions or the methods of analysis and to reaffirm the safety of their drilling technology. The questions raised, however, are serious enough to demand further study and rigorous scientific research.

As the much needed research into the risks associated with hydraulic fracking continue, it is wise to remember that neither the industry nor its critics have enough knowledge to provide the answers we need. It would certainly be premature to abandon natural gas as a viable energy option or horizontal fracking as a means of natural gas extraction. But it would also be premature and *utterly irresponsible* to proceed as though the industry's belief that its practices and technology are perfectly safe is an uncontestable truth. The research is incomplete to be sure, but enough has been done to recommend caution. The need for risk assessment and identification of risk management techniques is clearly required as an urgent necessity.

The concerns related to public health and safety in relation to fracking risks cannot be left in the hands of the energy producers and their supporters alone. Their first priority is clearly and understandably their own economic self-interest. That's business (and politics) as usual. Safety and public health are concerns that require an active governmental role and the application of the best scientific research. What is required is the performing of risk assessment as a public function and the promotion of risk management as a policy priority. With respect to hydraulic fracking and natural gas extraction, as well as with most technical advances and new technologies that expand risks to public health and safety, the more efficient and effective application of risk assessment and risk management techniques must become a prerequisite for policymaking and a foundation for identifying and establishing any needed regulatory requirements. But government has been either inconsistent or, in the case of the federal government, absent altogether with respect to the assessment of risks associated with fracking.

The 2005 Energy Policy Act in essence created a policy void removing the EPA as a monitor on the boom in gas fracking. This void was not, as the natural gas industry suggests, adequately filled by the states. State efforts vary greatly and, in most cases, are heavily influenced by the industry. Some states, Colorado for example, are said to provide reasonable protection for their residents from groundwater contamination. The State of New York, after discovering untreated fracking wastewater in its bodies of water, suspended fracking in the Marcellus Shale as it set about to create new protection rules (Climate Progress, 2011). But most states, especially those new to the fracking revolution, have few if any safeguards. Nevertheless, the fracking revolution has spread, and the risks associated with the technology have become a matter for wider public concern and discussion. Some states and now even the federal government are doing what should have been done long ago. They are getting serious about risk assessment and risk management. The policy void may soon be filled.

The EPA's 2010 decision to reconsider its 2004 assessment and launch an intensive study to learn whether the technology associated with the fracking boom in natural gas production is a threat to drinking water and to public health is an important first step on the path to responsibility. But both the EPA and Congress should have taken this step at the beginning of the boom. Having taken this step now does signal the beginning, however belatedly, of a necessary process for risk assessment. As the EPA completes its study over the next several years, it should also determine whether the urgency of the concerns that have led to this decision to study might require some regulation of hydraulic fracturing even before the assessment is completed. Several options are obvious.

The EPA could delay any regulatory recommendations or decisions about hydraulic fracturing until its study is completed. This would leave the question of regulation up to the states, as is currently the case, and allow the status quo to prevail. As a second option, the EPA could place a moratorium on all fracking operations until its study is completed and new federal regulations are developed. A third option would be to begin, based on evidence already available in existing studies and the concerns expressed or being addressed at the state level, to regulate hydraulic fracturing immediately in a manner that balances the need to protect public health and safety with the benefits gained from hydraulic fracturing. This third option has gained some support in the U.S. Congress but not enough to result in new legislation.

An effort to legislate and address the concerns associated with hydraulic fracturing was finally initiated in Congress in 2009, almost a full decade into the fracking boom. Those efforts failed but were renewed in 2011 with the reintroduction of the Fracturing Responsibility and Awareness of Chemicals Act (FRAC Act). The Senate version of this bill would close the oversight gap that the natural gas industry has benefited from since the passage of the 2005 Energy Policy Act. It would repeal the provision of the 2005 Act that exempted the industry from complying with the Safe Drinking Water Act. The bill would also require the public disclosure of chemicals used by the natural gas industry in its fracking operations, although they would not be required to reveal specific formulas where there is a proprietary interest. However, there is an emergency provision that would require proprietary chemical formulas be released to attending physicians, the state, and the EPA where the information is needed for treatment in emergency situations (S. 1215, 2011).

The FRAC Act died in committee in 2009. Its reintroduction in 2011 did not lead to passage either. The natural gas industry is lobbying aggressively against this legislation as part of its overall agenda to limit federal oversight of gas drilling. Congress remains far from united in perceiving the need to act. It is clear that the process of risk assessment and risk management will most likely continue to be pursued in the American policy process in an overtly political manner, and, as such, it is far from efficient in serving the public interest in safety and health. A responsible approach to risk assessment and risk management is perhaps compromised to the degree it is dependent on the vagaries of partisan politics and economic self-interest. But the potential for responsible action exists nonetheless as both the public interest in and the new documentation of risks are beginning to place new demands before policymakers.

Just as the policymaking process (and the politics that influences it) may promote a lag between new technologies and the analysis of risks they may impose, the inevitable impact of those risks soon becomes an impetus to bring the policy process back to dealing with them. This inevitable pivoting of the policy agenda, a response to the growing perception of or experience with the risks posed by a new technology, opens up new possibilities that the policy process will in turn investigate. Policymakers as well as some natural gas producers have expressed a willingness to cooperate in determining the effects of hydraulic fracturing and establishing appropriate safe guards (Chesapeake Energy, 2011). In a policy environment where risk assessment and risk management was a first and nonpartisan priority, this willingness could be tested and so much more efficiently capitalized on to promote industry cooperation in the thorough and credible analysis of the impacts that a surge in natural gas production may have on air, water, and landscapes. It could lead to expanded industry efforts to reduce methane releases during the production and distribution of natural gas, the establishment and regulatory enforcement of best practices, and the public disclosure of toxic chemicals used in natural gas production.

As the policy discussion pivots to focus more on risk assessment and risk management, the challenge will be to create policies and regulations that are grounded in scientific understanding and achieve effective communication of fact-based assessments of environmental impacts. More science must be injected into the fracking conversation so policymakers will have the foundation they need for responsible action. As this reorientation or pivoting of the hydraulic fracturing discussion begins, it may also be useful to use the occasion of this reorientation to broaden the discussion to include ways of addressing the need for general improvement in the public functions of risk assessment and risk management in relation to technology. Well, one can dream.

As the policymakers dawdle and the policy discussion is slowly engaging the discussion of risks related to fracking, local emergency managers are already being presented with new risks to manage. Perhaps the most immediate hazards that fracking communities and their emergency managers face are associated with transporting millions of gallons of water and thousands of gallons of chemicals to each well site. A tanker spill can be devastating to rivers, streams, and the health of nearby populations. The number of tankers in and out of fracking sites, twenty-four hours per day no less, escalates the risk of such spills beyond anything that may have been previously considered in a particular community. Other hazard concerns observed and requiring special attention in fracking communities include or are related to storing the wastewater and chemicals that return to the surface during the fracking process, naturally occurring toxic chemicals that may also surface during the fracking process, moving and treating wastewaters, and managing air pollutants. The potential for methane leaks into well water and the threats associated with such leaks (e.g., explosion) also present hazards that local responders, managers, and communities must take into consideration. Likewise, there will be new public health concerns associated with respect to exposure to chemicals and toxic wastes that may be visited on residents or workers at drilling sites. Some of these concerns are just beginning to be understood in fact. These are the obvious and immediate concerns, but others less obvious are beginning to emerge as well.

Recent studies have suggested that fracking may be related to increased risk of earthquakes. It is not actually fracking but the injection of the wastewater into disposal wells that seems to be the problem. Fracking involves, as we have seen, the high-pressure injection of fluids (water and chemicals) and sand into the ground to open up cracks in shale rock so that natural gas may escape and be captured. This, according to some recent studies, lowers the barriers to earthquakes (i.e., it loosens up the rocks enough to make an earthquake more likely). In Arkansas, it was discovered that wastewater disposal wells (i.e., high-pressure injection of wastewater into dead wells) were associated with an increase in earthquake activity. The number of incidents diminished quickly back to normal levels when wastewater wells were shut down (Shahan, 2011). In fairness, there have been subsequent studies indicating that the risk of earthquake in relation to fracking is minimal, but it is a risk nonetheless. Of course, we are at the beginning of the analysis, and whether in relationship to earthquakes or any other hazard risk, the cumulative impact of fracking activities (over say the next twenty or fifty years) may dramatically alter any of our current notions about hazard risks associated with horizontal fracking. One would suspect, based on some of the preliminary study to date, that these risks will be greater than we presently know. The continued study and refinement of our understanding of both the immediate and cumulative impacts of fracking on the environment, the water supply, the air, the health, and the safety of our communities should be regarded as an urgent necessity and not an inconvenience imposed by people or groups opposed to fracking. It is, rather, a vital component in the smart and necessary forward thinking and planning that creates and maintains sustainable hazard-resilient communities.

Whether drillers or emergency managers take seriously the threats posed, potentially at least, by fracking, it is interesting to note that one major insurance company has taken them very seriously. In the summer of 2012, Nationwide Mutual Insurance Company became the first insurance company to publicly state it won't cover damages caused by hydraulic fracturing. Nationwide said that its policies were not designed to cover the risks posed by fracking and that the risks associated with fracking operations were too great for them to ignore. The potential for their impact was considered too risky for them to cover (Speer, 2012). It is certainly advisable, and at the very least prudent, for emergency managers in fracking communities to take any hazard risks associated with fracking seriously and to factor them into their assessments of hazard risk, vulnerability, resource needs, and response capacities.

THE FUTURE OF EMERGENCY MANAGEMENT

It could be any community USA. A minor explosion caused by a mixture of swimming pool chemicals in a container forces residents and pets out of their luxury apartment complex. Fire department, police, and local HAZMAT team respond quickly and efficiently. They know the building, they know its systems, there is no wasted motion, and within two hours, the air is completely cleared, and the residents and pets are allowed back into their homes. All are amazed at the efficiency of the first responders and at the concern they showed for residents keeping them safe and informed. They even demonstrated concern for the pets. It's just another day at the office for the first responders.

It could be any small town or mid-sized city USA. A train derailment and chemical spill or an explosion at a local chemical plant brings first responders into action. Fire department, police, and HAZMAT teams respond quickly and efficiently. Residents are evacuated from harm's way. The first responders know the situation, they are prepared to handle it, there is no wasted motion, and within a reasonable time, the event is over, the area is cleared, and residents are safely back in their homes. All are amazed at the efficiency of the first responders and at the concern they showed for residents and their efforts to keep them safe and informed. Again, it is just another day at the office for first responders.

There are a considerable number of chemical- or industrial-related hazards that may produce disasters, and emergency management practitioners and first responders do an excellent job of responding to them. These hazard threats vary from location to location, but each community must be prepared to deal with them. Hazard resilience and sustainable community development require that forward thinking with respect to these potential hazards, proactive planning to mitigate and reduce the risks and vulnerabilities associated with them, and the development of capacities to respond and contain the damage should the hazard result in a disaster incident must all be an integral part of community decision making. It is well understood and accepted that our technical and industrial advancement does in fact create new hazards that we must be prepared to address. It is also well understood and accepted that progress requires a certain willingness to take risks. But these risks must be taken intelligently, and the management of them (i.e., minimizing their negative impact, reducing vulnerabilities, promoting hazard resilience and safety, etc.) is a necessity for sustainable development. Practicing emergency managers as a part of the community development network and the emergency management profession generally play a critical role in this regard, but both the practitioner and the putative profession must be more acutely aware that the hazards we produce, and the risks attendant to them, often exceed our immediate capacity or willingness as a society to manage. That may be the ultimate conclusion to be derived from the two specific cases we have discussed in this chapter.

An article appearing in the *Boston College Environmental Law Review* developed two accounts of why the risks of technological failure and the potential for disaster were spectacularly ignored by regulators and industry alike in the Deepwater Horizon case (Barsa and Dana, 2011). The authors argued that the inattention to risk was the product of a "group-think" pathology within the industry and among regulators. This involved an uncritical orthodoxy in which the safety of deep water drilling came not just to be accepted but was required within both groups. More compelling an explanation perhaps, and one consistent with our discussion of both Deepwater Horizon and hydraulic fracturing (fracking), the second account concludes that the inattention to risk by the industry was a rational decision based on the benefit to the industry in cost reduction. This objective was in no small measure aided by the industry's ability to capture regulators to avoid unwanted scrutiny (Barsa and Dana, 2011) and, it might be added, the reluctance of lawmakers to regulate.

Both of the cases we have examined demonstrate the reluctance of industry to acknowledge the risks associated with their activities. The avoidance of investments in safety, and the creation and implementation of any governmental regulations that might require it, is a common tactic incentivized by the goal of reducing costs and maximizing profits. It seems, considering our discussion of both the Deepwater Horizon and fracking, that industries are increasingly successful at creating new hazards, introducing disaster risks, and negatively impacting the hazard resilience and sustainability of communities without bearing any of the costs or assuming any responsibility. It is impractical to think that emergency managers and first responders, no matter how dedicated and well prepared, can keep pace with the new hazards and risks this creates. Without the creation and enforcement of mitigation and safety requirements that make the private sector a partner with the public sector in hazard mitigation, hazard resilience may be impossible. As we noted, risks cannot be left in the hands of the hazard creators (i.e., private industry) and their supporters alone. Their first priority is clearly and understandably their own economic self-interest. Absent incentives to do otherwise, they will reduce costs by rolling the dice on risk. Safety and public health, not to mention sustainability and hazard resilience, are concerns that require an active governmental role and the application of the best scientific research. What is required is the performing of risk assessment as a public function and the promotion of risk management as a policy priority. Unless that partnership is created and enforced, and it won't be without the performance of and promotion of risk assessment as a public policy priority, local emergency managers and first responders will be unable to keep pace with the creation of new hazards and the introduction of new risks.

With respect to a disaster such as Deepwater Horizon, the lessons to be learned suggest the need to improve risk assessment as a basic first step. Given the lax attitude noted in that case toward investment in safety and the propensity to elevate cost cutting to the level of highest priority, it is not unreasonable to suggest that policymakers take steps to require enhanced risk assessment and baseline procedures and guidance for safety preparedness efforts. The verification of safety preparedness (i.e., verification by independent review), makes sense as well. But how likely is this to be considered? The reality of persistent and largely successful industry and private sector resistance to any regulation, and a political policymaking environment more sympathetic to the hazard creators than to the emergency manager, means that the goal of hazard resilience often runs against the grain of our economic and political systems. Emergency managers, and the putative emergency management profession, must be intelligent, forceful, and full participants in expanding the national dialogue to promote a different reality.

At a minimum, in an effort to propel risk assessment and hazard mitigation to a priority position within the policy process as it relates to new technologies that create hazards and carry potential risks with them, three general principles should be elevated above all other competing and valid concerns and interests. First, the policy discussion must be infused from the beginning with the appropriate and necessary scientific foundation. The development of the necessary foundation for knowledge-based decision making must not be short-changed in the more immediate service of any other political, economic, or ideological interest, however valid these may be. Second, as the science is infused, it must be directed to promote a comprehensive risk assessment as a prerequisite for any long-term policy development. Third, the implementation of new technologies for production proven to be associated with the introduction of hazard risks should be subject to reasonable and scientifically legitimate requirements for mitigation planning and disaster or emergency preparedness planning.

The application of these three principles to fracking, for example (i.e., at the beginning of the development and implementation of horizontal drilling techniques), would have looked like what appears to be happening belatedly some ten years into the new boom in natural gas production. The EPA research program to access the environmental and hazard risks associated with fracking would have been pursued as a first step in the policy process and completed, or at least advanced, years earlier. The exemption of natural gas producers from federal regulation (i.e., the Safe Drinking Water Act, the Clean Water Act, and the Clean Air Act) would not have been enacted prior to the scientific study to inform the policy. The void created by removing the EPA as a monitor on the boom in gas fracking, a boom that has cut across state lines into almost every region of the country with the potential to impact public health, safety, and sustainable hazard resilience across the country, would not have been allowed to exist. The legitimate concerns related to the contamination of drinking water, the toxicity of fracking chemicals and waste waters, the correlation of drilling and high methane levels in ground wells, and a host of other concerns would have been much more thoroughly explored and much earlier. Responsible interim legislation, perhaps something like that proposed in the

FRAC Act, would have been enacted years ago. By now, some ten years into the fracking boom, the foundation for knowledge-based decision making would have led to the establishment of sensible federal and state regulations protecting public health, mitigating hazards, promoting safety, and ensuring no doubt the sustainable future of shale gas extraction. From the very beginning, policymakers and the natural gas industry would, through the determined application of these three principles, have been full partners in the tasks of hazard identification, vulnerability assessment, and disaster mitigation. As we have said, it does take a village to promote sustainable and hazard-resilient communities. But as our two cases suggest, there is no village when those who introduce the hazards are not required to participate in managing the risks and when those who create the policies are reluctant to require that they do so.

There is no doubt, as we have noted repeatedly, that technological and economic progress or advancement requires risks to be taken. New human-created hazards, will accompany new initiatives. There is also no question that the industries creating those hazards and taking the risks and the public policymakers who monitor and potentially regulate them are interested in a number of valid and competing economic and political objectives. However, given that major and complex technological advances carry the potential for creating new hazards and inflicting greater risks on our communities and their populations, it follows that risk assessment and risk management must be higher priorities for public policymakers. It follows also that those whose trade or profession is the assessment and management of hazard risks and vulnerabilities should be at the front of the chorus of voices promoting forward-looking and reasonable policies, regulations, and practices that address the human-made hazards and the risks that inevitably come as a byproduct of progress.

The fact of the matter is it is hard to promote a proactive approach to the identification of hazards, risks, and the mitigation of their threats. The two examples we have discussed are no different than numerous others we might examine. Economic incentives and political expediency, not to mention the vagaries of partisan politics, work against the forward thinking required for risk identification and hazard mitigation. All of the excellent work that practicing emergency managers may accomplish with regard to mitigation and hazard resilience can be swiftly overshadowed by the introduction of new hazards and risks associated with unregulated or underregulated industrial advances. The political and policy process tends to be reactive, not proactive, with respect to hazards and risks. The general public tends to be reactive as well. The private sector, as our two cases suggest, tend to deny risks or be indifferent to them until a disaster is experienced. Keeping production costs down is, invariably, a more important goal in the scheme of things, and sacrifices in safety and risk management are all too commonly incentivized by management as a result.

It may be too idealistic to suggest that the public policy process and the general public will ever be anything more than reactive with respect to the assessment and management of human-made risks associated with new technologies. It may be too idealistic to suggest that industries and businesses will ever change their incentives and place a greater and more scientific focus on hazard mitigation, public safety, and disaster prevention. But it should not be too idealistic to expect that practicing emergency managers and the putative emergency management profession should promote a more proactive and responsible approach in the policy dialogue and in their relationships at all levels (national, state, local community) with the private sector. When it comes to hazard mitigation (i.e., the forward thinking that sustainable and resilient communities must do in relation to hazards and risks), emergency management may be the only adult in the room. The training, work, and experience of its practitioners and the perspective of the profession place it in the position, however uncomfortable or undesirable it may seem to some, of encouraging proactive and responsible action from others. It takes a village, but somebody has to educate the village and encourage it to act.

Better mechanisms for the promotion of proactive mitigation and risk reduction strategies are needed to be sure. Reactive responses, or the simple denial or ignoring of new hazards and risks in the pursuit of immediate economic benefits, only postpone the inevitable at the expense of health, safety, sustainability, and community resilience. Emergency management must, beyond its considerable and excellent technical competencies, lead the conversation. As a sustainability profession, if indeed that is what it might be said to be, emergency management must take the long view, must be proactive, and must demonstrate in as compelling a fashion as possible the costs and risks of the short view. It must stimulate broader thinking and discussion about the hazards we create and the risks we must manage as a result of our own creation. In so doing, it must also know, as Pogo knows, who the enemy is. It is absolutely us. Preventing the next Deepwater Horizon disaster will not be possible until we learn to stop harming ourselves and our environment by fighting against hazard resilience for the sake of profit or progress. That's a net loss for everybody.

Communities are diverse, multilayered, and interconnected. They each chart their own course with respect to economic and community development. Communities in charting their path may, without intending to do so, endanger their citizens and natural assets. Hazard resilience, especially in relation to the risks produced by our own actions and designs, is not something that happens naturally in the course of human events. It is something to be carefully planned and implemented and, as such, requires someone to take responsibility. Responsibility for hazard mitigation, for the most part, is the job of emergency managers. They work with communities to build mitigation capacities and protect them from both the hazard threats posed by nature and the excesses of their own hazard-creating activities. Neither practicing emergency managers nor the putative profession must forget that a large part of their work is protecting us against ourselves.

REFERENCES

Barsa, M., & Dana, D. A. (2011). Reconceptualizing NEPA to Avoid the Next Preventable Disaster. *Boston College Environmental Law Review, 38*(2), 219–246.

Beder, S. (2002). *Global Spin.* United Kingdom: Green Books.

Bolstad, E., Goodman, J., & Taylor, M. (2010). After Argument, BP Official Made Fatal Decision on Drilling. Available at http://www.mcclatchydc.com/2010/05/26/94859/after-long-argument-bp-official.html. Accessed June 1, 2010.

Botkin, D. B. (2010). *Powering the Future.* Upper Saddle River, NJ: Pearson Education.

Bruzelius, N. (2010). EPA Turnaround: Collecting Data on Fracking Risks Might Be a Good Idea. Available at http://www.ewg.prg/kid-safe-chemicals-act-blog/2010/03/epa-turnaround-collecting-data-on-fracking-risks-just-might-be-a-good-idea/. Accessed May 25, 2011.

Chesapeake Energy. (2011). Commitment to Environmental Excellence. Available at http://www.chk.com/Environmental/Commitment/Pages/information.aspx. Accessed August 3, 2011.

Climate Progress. (2011). Getting to the Bottom of Natural Gas Fracking. Available at http://www.thinkprogress.org/romm/2010/03/03/205585/natural-gas-frackin/. Accessed December 5, 2011.

Earth Works Action. (2011). Hydraulic Fracturing 101. Available at http://www.earth-worksaction.org/issues/detail/hydraulic_fracturing_101. Accessed December 5, 2011.

Energy and Commerce Committee. (2010, May 25). Memorandum from Henry A. Waxman (Chair) and Rep. Bart Stupak (Ranking Minority Member) to Members of the Subcommittee on Oversight and Investigations. Washington, DC: U.S. House of Representatives.

Energy and Environment Subcommittee. (2010, July 12). Hearing to Explore Technologies, Standards, and Practices. Washington, DC: U.S. House of Representatives.

Gallant, B. J. (2008). *Essentials in Emergency Management.* Lanham, MD: Scarecrow Press.

Hanson, K. (2010). Safety, Corporate Responsibility, and the Oil Spill. Available at http://www.scu.edu/ethics/practicing/focusareas/business/bp.html. Accessed December 5, 2011.

Hartenstein, M. (2010, June 24). BP Oil Leak by the Numbers. *New York Daily News.*

Howarth, R. W., Santoro, R., & Ingraffea, A. (2011). Methane and the Green-House-Gas Footprint of Natural Gas From Shale Formations. *Climate Change, 106,* 679–690.

Jamarillo, P., Griffin, W. M., & Mathews, H. S. (2007). Comparative Life-Cycle Air Emissions of Coal, Domestic Natural Gas, LNG, and SNG for Electricity Generation. *Environmental Science and Technology, 41,* 6290–6296.

Johnson, B. (2011). Official Investigation: BP's Risky Efforts to Cut Costs Caused the Deepwater Horizon Disaster. *Think Progress.* Available at http://thinkprogress. org/climate/2011/09/14/318898/official-investigation-bps-risky-efforts-to-cut-costs-caused-the-deepwater-horizon-disaster/?mobile=nc. Accessed July 22, 2012.

Katusa, M. (2011). The Fracking Controversy. Financial Sense. Available at http://www.investorsinsight.com/blogs/essay_research/archive/2011/05/19/the-fracking/controversy.aspx. Accessed December 5, 2011.

Kirby, A. (1990). *Nothing to Fear: Risks and Hazards in American Society.* Tucson, AZ: University of Arizona Press.

Lindheim, J. (1989). Restoring the Image of the Chemical Industry. *Chemistry and Industry, 15,* 491–494.

Maugeri, L. (2010). *Beyond the Age of Oil.* Santa Barbara, CA: Praeger.

McClatchy, D. C. (2010). BP Has a Long Record of Legal Violations. Available at http://www.mcclatchydc.com/2010/05/08/93779/bp-has-a-long-record-of-legal.html#ixzzOnAzTzTzdgN. Accessed June 28, 2010.

Merrill, R. (1991). Activism in the 90's: Changing Rules for Public Relations. *Public Relations Quarterly, 36*(3), 28–32.

Mitchell, J. K. (1996). Improving Community Responses to Industrial Disasters. In J. K. Mitchell (Ed.), *The Long Road to Recovery.* New York: United Nations Press.

Osborne, S. G., Vengosh, A., Warner, R. W., & Jackson, R. B. (2011). Methane contamination of drinking water accompanying gas-well drilling and hydraulic fracturing. *Proceedings of the National Academy of Sciences.*

Ott, R. (2005). *Sound Truth and Corporate Myth: The Legacy of the Exxon Valdez Oil Spill.* Cordova, AK: Dragonfly Sisters Press.

Revesz, K. M., Breen, K. J., Baldassare, A. J., & Burruss, R. C. (2010). Carbon and Hydrogen Isotopic Evidence for the Origin of Combustible Gasses in Water Supply Wells in North-Central Pennsylvania. *Applications in Geochemistry, 25,* 1845–1859.

Shahan, Z. (2011). Oklahoma Earthquakes and Fracking. Available at http://planetsave.com/2011/11/07/oklahoma-earthquake-fracking/. Accessed July 24, 2012.

Shindell, D. T., Faluvegi, G., Koch, D. M., Schmidt, G. A., Unger, N., & Bauer, S. E. (2009). Improved Attribution of Climate Forcing to Emissions. *Science, 326,* 716–718.

Shires, T. M., Loughran, C. J., Jones, S., & Hopkins, E. (2009). Compendium of Greenhouse Gas Emissions Methodologies for the Oil and Natural Gas Industry. Prepared by URS Corporation for the American Petroleum Institute, Washington, DC.

Skrapits, E. (2011). Environmental Watchdog Outlines Fracking Risk. Citizensvoice.com. Available at http://www.citizensvoce.com/news/drilling/environmental-watchdog-outlines-fracking-risks-1.1115694#azz1JR6rMJa1. Accessed April 14, 2011.

Southern Studies. (2010). Oil Spill Reveals Dangers of Deep Water Drilling. Available at http://www.southernstudies.org/2010/06/oil-spill-reveals-dangers-of-deepwater-drilling.html. Accessed July 12, 2010.

Speer, M. (2012). Fracking Risks Are Too High for Nationwide Insurance. Available at http://www.isustainableearth.com/energyefficiency/fracking-risks-are-too-great-for-nationwide-insurance. Accessed July 14, 2012.

Strong, C. B., & Irwin, T. R. (1996). *Emergency Response and Hazardous Chemical Management.* Delray Beach, FL: St. Lucie Press.

Thomas, P. (2010). Offshore Drilling: Years of Lax Oversight, Small Fines for Serious Violations. Available at http://abcnews.go.com/WN/oil-spill-mms-offshore-drilling-regulation-small-fines/story?id=11003043. Accessed July 22, 2012.

Urbina, I. (2010, May 24). Ethical Lapses: The BP Oil Spill. *New York Times.*

Vaughan, D. (1996). *The Challenger Launch Decision.* Chicago, IL: University of Chicago Press.

Wakatsuki, Y., & Mullen, J. (2012). Japanese Parliament Report: Fukushima Nuclear Crisis Was Man-made. CNN.com. Available at http://www.cnn.com/2012/07/05/world/asia/japan-fukushima-report/index.html. Accessed July 5, 2012.

Zoback, M., Kitasei, S., & Copithorne, B. (2010). Addressing the Environmental Risks From Shale Gas Development. Worldwatch Institute Briefing Paper 1. Available at http://blogs.worldwatch.org/revolt/wp-content/uploads/2010/07/Environmental-Risks-Paper-July-2010-FOR-PRINT.pdf.

Chapter 6

RESILIENCE AND
SUSTAINABILITY BY DESIGN

Floods are 'acts of God,' but flood losses are largely acts of man.
— Gilbert F. White

A BASIC INSIGHT

Gilbert F. White (1911–2006), a prominent American geographer whose work spanned almost seventy years, is known as the father of floodplain management. Beginning with his doctoral dissertation in 1942, White promoted the comprehensive management of floodplains through the adoption of a broad range of adjustments to floods. These included adaptation to or accommodation of flood hazards rather than the reliance solely on the structural solutions (i.e., dams and levees) that dominated flood policy in the early twentieth century.

As he studied the recurrent Mississippi River floods, White challenged the notion that natural hazards were best controlled by engineering and improved construction techniques. While improved engineering and construction techniques can be useful to mitigate natural hazards in the short term, White's insight was that these often postponed disaster damages rather than preventing them. In the long term, structural mitigation might even increase the severity of flood damages. His research demonstrated that flood control structures, for example, not only frequently failed to meet the standards of reliability set by planners but often led to increased damages in the long term. Structural solutions might increase damages to the extent that they lull a community to invest unwisely in the development of high-risk locations under the false assumption that the land could be permanently protected through innovative structural design alone. White thus emphasized the need for nonstructural solutions, such as zoning restrictions and flood-proofing requirements, to complement or, where feasible, replace the more traditional structural approaches.

White's point of view was considered quite radical when first articulated. But over a fifty-year period, his radical insight became common wisdom. Land-use planners, scientists, and governments around the world today look at the landscape the way Gilbert White did beginning in 1942. They balance a range of alternatives that include upstream watershed treatment, the flood-proofing of buildings, emergency evacuation procedures, dams, and land use policies. His legacy among geographers is of course unrivaled. But in the broadest sense, and beyond his contributions to floodplain management, White promoted environmental stewardship. Preserving nature and promoting the sustainable use of the earth's resources were necessities in his mind as a primary means of managing the threats posed by natural hazards and ensuring that our communities would be both resilient and sustainable. White's interests and work thus included the study and mitigation of a wide range of natural hazards. By the 1970s, for example, his attention was drawn to the increasing loads of greenhouse gases in the atmosphere. He began to worry that human activity might cause a change in the global atmosphere and contribute to some dire consequences with respect to sustainability.

White's basic insight, one might suggest, is one with which all practitioners in the field of emergency management must begin their life's work. It is, in fact, the basic insight that might well express the need for and the ultimate purpose of an emergency management profession. Aptly articulated in the quote with which this chapter begins, "floods are 'acts of God,' but flood losses are largely acts of man," White's basic insight is that human beings must accept responsibility for disasters or, more precisely, for the damages they may cause. Human beings, and not nature, are the primary cause of disaster damages. The choices we make about where to live, how to live, and how human and economic development will proceed determine the losses that we will bear in future disaster scenarios. The choices we make today will shape the nature, impact, and severity of the disasters that will inevitably occur tomorrow in the shadow of the hazards that present themselves to us and our communities in the natural course of events or that are spun off from human technological advancement. This basic insight is confirmed by science and experience, as are our limitations in managing the threats that hazards may pose.

White's insight includes an awareness born of hard experience and scientific study that structural innovations alone are not sufficient to adapt to natural hazards or, we might add, to human-created hazards. Technology has its limitations, and it cannot, despite our faith in it, make our communities entirely safe from all forces of nature. Structural innovation has its utility, but its limits cannot be ignored. The sustainability of our communities also requires that we understand, respect, and protect the natural order by intelligently accommodating ourselves to its inevitable extremes. A failure to do so, or the choice to refrain from doing so, only results in disaster losses that are the product of human design.

DISASTERS BY DESIGN

Emergency managers are of course not really strangers to White's basic insight. In fact, that basic insight is a theme that has been running throughout this book and throughout the work of emergency managers for decades. For more than twenty years, since the beginning of the mitigation decade of the 1990s (Chapter 2), emergency managers have been trained to look at the world differently. They have come to understand that natural disasters, for example, are not entirely the work of nature. As we said in Chapter 1, they are the predictable results of interactions among the earth's physical systems or natural environment; the social, economic, and demographic characteristics of the human communities in that environment; and the specific features of the constructed environment (buildings, roads, bridges, and other features). Emergency managers have learned and put into practice mitigation techniques that may reduce the impact of natural hazards. As noted in Chapter 2, hazard mitigation is not only one of the four disaster phases that trained practitioners must master, it is perhaps the most critical phase with respect to the sustainability of our communities. Hazard resilience is an absolute necessity for a community to be sustainable. The linkage of hazard mitigation, as an emphasis or priority in the emergency management cycle, to the broader task of developing sustainable communities has placed emergency management at the heart of community development and planning. Emergency managers must comprehend that as a basic prerequisite for understanding their role and, it will be suggested, defining their profession.

Emergency managers and the communities they serve must take responsibility for disaster losses (i.e., see them as "acts of man"). They must understand the three major influences (i.e., the earth's physical or natural systems, the demographic characteristics of the human community, and the built or constructed environment) that contribute to the level of disaster damages. These influences, identified as the roots of the problem in Dennis Mileti's classic book, *Disasters by Design,* and an understanding of them make it abundantly clear that disasters are not unexpected events, and many, if not most, of the losses associated with them are predictable as well as preventable. The first step toward creating hazard-resilient communities, a necessary step to building and maintaining sustainable communities, and the first priority of a putative emergency management profession must be to *stop designing disasters.* This is easier said than done of course. As we have seen, whether with respect to natural or human-made hazards, such progress is easily offset by the shorter-term goals and objectives that animate so much of our economic and political lives. We seem to have a hard time remembering what we have learned or what we know, by both experience and by science, as we pursue what we want.

The physical environment is constantly changing. Being aware of and designing our communities so that they consider those changes, act in harmony with them to some extent, is a practical and necessary thing. As we saw in our discussion of climate change (Chapter 4), the warming of the climate is projected to result in more dramatic meteorological events. Storms, floods, extreme temperatures, drought, wildfires, and other natural hazards will be altered accordingly. These changes suggest that the nature and course of even those hazards that may be anticipated on a routine or regularly recurring basis in every community are subject to changes with respect to frequency, severity, or intensity. The past is no longer a reliable source of information as the impact of climate change will change the vulnerability profiles of communities as it changes the course of the natural hazards it experiences on a repeating basis. Not only must emergency managers be forward looking in anticipating these changes, but the broader task of managing the risks must include efforts to mitigate future damages through the adoption of strategies that may address their causes.

It is inconceivable that either sustainable communities or emergency managers would not see the necessity of taking a significantly changing climate into consideration. All communities, and their practicing emergency managers, must take into consideration climate change projections relevant to their specific communities that will affect their identification and selection of hazard mitigation strategies, the preparedness activities that their jurisdictions will need to undertake, the nature and scope of their response operations, and the identification and implementation of appropriate recovery strategies that will contribute to resilience and sustainability. Emergency managers, as we emphasized in Chapter 4, must enhance the effectiveness of their traditional functions by incorporating climate change adaptation into their everyday work. Beyond all of this, mitigation in the form of reducing greenhouse gas emissions would be a logical and necessary step to reduce the number of disasters by design. That of course requires action by policymakers. But it is of course by no means inevitable that policymakers will act. Likewise, beyond changes in the physical environment, other changes require us to take note and respond.

The demographic composition of the United States and changes in it also contribute to the creation of new disaster-related risks and costs. The number of people living in earthquake-prone regions, flood zones, and in coastal areas subject to hurricanes is growing rapidly. Over one-half of the U.S. population, for example, lives within fifty miles of a coast. The high-density development of coastal regions, on roughly 17% of the nation's land, exposes millions of more people and large amounts of upscale development in the path of hurricanes, thus contributing to the rising costs of disasters. Population density is also a concern. Eighty percent of Americans, according to the 2010 census, live in urban areas. This creates, along with the impact of the development strategies necessary to accommodate a larger number of people, new human and economic costs associated with disaster occurrence in these areas. The worsening inequality of wealth, a growing phenomenon in the United States over the past several decades, also contributes to disaster costs. It not only may make more people vulnerable to hazards, but it generally visits the worst impacts and greatest costs on those who are less able to recover from them on their own (i.e., the poor). According to the 2010 U.S. Census, pockets of extreme poverty in the United States increased by 41% from 2000–2010. An awareness of these demographic trends and their meaning with respect to the changing vulnerabilities of populations and the potential costs of disasters is critical to emergency managers. Reducing the vulnerabilities and costs is what sustainable hazard mitigation is about and, one would presume, should be a priority for all communities. This too will require action by policymakers. But it is by no means inevitable that these demographic trends will be addressed by them in a timely fashion.

The constructed environment (homes, businesses, transportation systems, public utilities, communications systems, etc.) is also an important factor in increasing the costs of disasters. The growing density of the population makes the potential damage to the accompanying constructed environment and losses from disasters larger and more expensive. But the worst impacts may be those resulting from the destruction of local ecosystems that provide protections from natural hazards. Again, it is not inevitable that we will stop destroying ecosystems as we build or extend our communities.

The draining of swamps or the bulldozing of hills often makes attractive and profitable development possible in high-risk locations, but it may also degrade the environment and expose the new development to heightened risks and the potential for greater losses in disaster scenarios. Even well-intended mitigation efforts to protect new developments may, in the long term, increase the damages and costs associated with natural disasters. As we saw in our discussion of New Orleans and Hurricane Katrina, the greatest damages associated with it may not have been the work of nature. They were, we said, as much the result of the acts of developers, speculators, engineers, and politicians who ignored the signs of disaster resident in unsustainable development practices. The structural protections (i.e., levees) were unable to prevent the damages, thus suggesting once again that structural or engineering solutions alone are often inadequate. These may postpone losses (e.g., levees that are built to provide flood protection may work today but fail tomorrow as circumstances change). This postpones damages perhaps for a significant time. It may create a sense of security and discourage doubt as to the adequacy of our engineered systems, but when the structures do fail, the amount of property damage and the human impacts are both more costly and more tragic.

As much as human actions and decisions in relation to earth's physical or natural systems, the demographic characteristics of the human community, and attributes of the built or constructed environment (the three roots of the problem in relation to natural disasters) may contribute to the "designing" of disasters and the rising cost of their damages, they are not alone in creating disasters. It is also quite obvious that the creation of new hazards and the taking of new risks in conjunction with technological and economic advancements is a potentially potent cause of disasters by human design.

The Fukushima Daiichi disaster, the Deepwater Horizon disaster, and the risks associated with hydraulic fracturing (all discussed in Chapter 5) are recent examples that indicate that nature is not the only source of hazards. Human invention, for all of its potential for good, includes as a byproduct the unfortunate and costly capacity to design, build, and even tolerate ever more devastating disasters that may threaten sustainability and degrade the natural environment. These examples demonstrate that corporate and political decision makers are quick to minimize risks attached to promising new technologies and practices. As the discussion in Chapter 5 demonstrated, they tend to overlook or underestimate or even to be willfully negligent with respect to the identification of long-term risks. They are often too inclined, as we saw in Fukushima, in the Gulf, and in the natural gas drilling activities underway across the United States, to neglect investments in safety as they pursue more immediate economic benefits and political objectives.

Whether with respect to natural hazards or the human-made hazards that accompany our technological development, emergency preparedness and disaster response efforts have worked hard to keep pace and deal with the destruction, losses, and human suffering imposed by disasters. But can the same be said with respect to hazard mitigation? Have we succeeded in doing everything we know to do to prevent disasters or reduce risks and disaster losses? If we have not, might that handicap even the best preparedness and response efforts as events race quickly beyond their reach?

As we saw in our discussion of mitigation in Chapter 2, significant steps have been made over recent decades to be more proactive about mitigating hazards in a sustainable way. Structural mitigation in particular has been embraced in many communities across the country. Examples of the retrofitting of buildings to withstand the blowing of high-velocity winds or the violent shaking of the ground are commonly observed, and their contribution to reducing disaster damages is well documented. The intelligent management of floodplains continues to reduce the damages and losses associated with flooding. The principles and techniques of sustainable hazard mitigation are no strangers to communities and emergency managers across the nation. Yet the effectiveness of hazard mitigation, of efforts to stop designing disasters if you will, is both inconsistent and troubling. Despite its effectiveness, hazard mitigation is not inevitable in any community.

The inconsistency of mitigation efforts is undoubtedly the product of a number of factors. First, national leadership has been inconsistent in promoting it as a national priority. Second, not every community has embraced it equally. Third, sustainable hazard mitigation is often overwhelmed by other "necessities" (e.g., economic development, job creation, energy independence, etc.) that are incorrectly seen as being in conflict with sustainability in general and hazard mitigation in particular. Fourth, and finally, our trust in and overreliance on structural mitigation measures have often led to underutilization and underprioritization of nonstructural measures. The troubling thing is that, for all of our observable and even admirable progress, we may actually be losing ground in the battle to take responsibility for disaster losses.

National leadership does matter. That was made clear in our comparison of the "mitigation decade" (the Clinton administration) and the "decade adrift" (the Bush administration) in Chapter 2. The merits of specific programs aside (e.g., Project Impact), it is clear that national priorities as articulated by national leadership make a difference. These priorities must, of course, be reflected in policy. Local communities respond to the initiatives and funding opportunities presented to them more frequently than they take the initiative to create new and innovative hazard mitigation programs on their own. Consistency of hazard mitigation efforts in communities across the nation is likely to be better achieved in the context of a national effort that effectively integrates and combines their efforts toward the same objectives even as it promotes flexibility, creativity, and local leadership.

Of equal if not greater concern than the ingredient of national leadership is the tendency to waiver in the commitment to sustainable hazard mitigation as a perceived priority or necessary thing in order to deal with other pressing and immediate problems. Public opinion is fickle. Changing circumstances, the debate among political elites, and media coverage may all influence it dramatically. Climate change is an issue where this is a prime example. American public opinion has bounced back and forth on this topic over the past decade. Between 2007 and 2010, a sharp decline occurred in the number of Americans who believed that climate change was real and needed to be addressed as a priority. That decline was attributed to the bad economy (the worry about jobs pushed climate and the environment to a lower priority), the debate among partisans (contributing to the doubt about the science or perception of the threat as exaggerated), and media coverage (or the lack of it). By the summer of 2012, public opinion polls showed a sharp increase in the number of Americans who believed that climate change was real and that policymakers should address it. The record summer heat and persistent drought experienced across much of the country were undoubtedly having an impact on public perceptions.

The fact of the matter is that immediate concern, the crisis de jour if you will, always seems to carry more weight in public conversations than the longer term objectives of sustainable hazard mitigation. Now, as in today, is never the right time to think long term or make changes in the way we live. Fixing a stalled economy takes precedence over sustainable development practices, our current energy needs dictate continued reliance on fossil fuels (i.e., the alternatives are not available, affordable, or practical in the short term), and of course the costs associated with sustainability enhancing practices are (always) declared prohibitive at the present time. The accepted reasons for not changing course and the unquestioning acceptance of them seem most frequently to prevail. Generally, this is to the detriment of a community and far more costly to it in the long run. But we do not live in the long run or so it seems sadly reasonable to conclude.

An assumption that the crisis de jour, whatever it may be, cannot be addressed while addressing longer term threats to sustainability is often the byproduct of short-term thinking. Short-term thinking, in turn, seems to be the human norm. This leads to the acceptance of what we are doing now as the necessary thing to do and, not infrequently, the assumption that the people doing what they are doing are the experts and the technologies that support their activities can be managed safely. These "experts" spend huge amounts of money to tell us everything they are doing is necessary, urgent, and safe. So we needn't worry, at least in the short term. We often accept as common sense the course we are on without questioning it. When the drillers for natural gas tell us that horizontal hydraulic fracturing is safe, we believe it because they have told us, and they are the experts. If we don't believe it entirely, we still may be willing to roll the dice a bit because we need the jobs or we are worried about America's energy future, and these things *must* take priority over our concerns about the safety of horizontal fracking and any risks it may impose on humans and the communities in which they live. The first wrong assumption (i.e., we have to deal with an immediate crisis first and postpone any concerns about sustainability in the long term) leads to a series of assumptions on which are built multiple disasters by design. This *does make inevitable* a dramatic increase and perhaps even an acceleration of disaster costs that are entirely of our own making.

Is it difficult to accept that humans make many wrong assumptions that lead, inevitably, to disasters? The unquestioning acceptance of nuclear power as an accepted and necessary thing by the Japanese, for example, an unquestioning acceptance that was said to be imbedded in its culture to such a degree that the nuclear industry became basically immune to careful and objective scrutiny, is not the exception, it is the rule. This is not to criticize the Japanese or any other nation. It is simply to state that the usual realm of human experience is short term (i.e., we live in the short term), and the assumptions (cultural and other) that govern the short term are such that it often takes disasters that expose human populations and social systems to consequences that fall outside the realm of most human experience to demonstrate the inadequacies of our present thinking. The Fukushima Daiichi disaster was in this sense inevitable. One may be tempted to say the disaster was also necessary to break through the layer of unquestioning acceptance that tolerated and even encouraged a tragic lack of critical thinking in the first place.

Sustainable hazard mitigation is a concept that was born not of the normal realm of short-term human experience but out of the changed reality imposed by the disasters and damages that were the product of human design in the short term. Dramatic events, we might say, are often necessary to change reality. With respect to natural and human-made disaster experiences, the changed reality produced an awareness that human thinking needed to change with respect to the way we viewed disasters. As James Lee Witt and FEMA in the 1990s demonstrated, reducing the repeated loss of property and lives every time a disaster struck was possible if we began to be forward in our thinking and proactive in our planning. But our thinking needed to take responsibility for disaster losses, expand its reach, and direct its efforts toward the requirements of sustainability and resilience as a first priority.

As we learned to think differently about disasters, we also learned that resilience and sustainability are not easy or inevitable choices that humanity makes. They are, to the extent they may be achievable, the result of thinking and acting differently. They are intended. They are a conscious choice. They too, like disasters and their costs, are the product of human design. Emergency managers are expected, of course, to be masters of anticipatory thinking. That implies that they cannot be indifferent to unsustainable practices or disasters by design. Indeed, they are among those who must exhibit leadership in designing resilience and sustainability.

RESILIENCE AND SUSTAINABILITY BY DESIGN

In Chapter 2, we emphasized that disaster or hazard mitigation begins with the realization that most disasters are not unexpected. This is to say, once again, that disasters are actually the predictable results of interactions among the physical or natural environment; the social, economic, and demographic characteristics of the human communities in that environment; and the specific features of the constructed environment (buildings, roads, bridges, and other features). This is to begin with the basic understanding, perhaps imperfectly articulated in the context of emergency management practice sometimes, that human action (or inaction) may contribute to the undermining of natural systems that are necessary to sustain a community into the future. Our human communities are a subsystem of and dependent on a larger but finite system, the biosphere. Over time, the capacity of the biosphere to provide essential ecosystems (clean air, water, arable land, oil reserves, etc.) will erode or decrease. The intensifying of this erosion by the growing demands of an expanding human population and human activities that place greater stress on ecosystems is, at its heart, a prime contributor to the incidence and costs of disasters. In other words, we might say, by way of example, that the growing per capita consumption rate of an expanding population combined with development strategies that do not sustain the natural environment contributes to the creation of severe hazard risks that endanger the future health of our communities and of the natural world. Disasters are, in this context, primarily sustainability issues (i.e., in large part, the damages they produce are a product of unsustainable practices).

It may be recalled that sustainability is generally agreed to mean the effective use of resources (natural, human, and technological) to meet today's needs while ensuring that these resources are available to meet future needs. Disasters, natural or technological, wreak economic and environmental havoc on human communities. They also have the unfortunate potential to accelerate the loss of ecosystems necessary to sustain human communities. That being the case, resistance and resilience in the face of predictable disasters is a critical necessity. The prevention of disasters, or at least the reduction of their costs to humanity and the environment, is among the critical characteristics of a livable and sustainable community. This requires more than structural adjustments that enable our human communities to sustain disaster impacts and recover from them quickly. It also requires the avoidance of activities that constitute direct threats to sustainability and the promotion of human development and living strategies that preserve the socioecological system. This involves the awareness that hazard resilience has its greatest value or worth insofar as it contributes to sustainability. It also requires a basic acceptance and understanding that resilience and sustainability are in no small measure the products of human design. Taking responsibility for disasters and their costs makes sense only in this context. Taking responsibility requires the designing of resilient and sustainable communities.

It is perhaps arguably true that we are in fact living at a time when society generally has placed a greater emphasis on balancing environmental and developmental concerns. Promoting livable communities and securing high-quality "green" development are increasingly popular themes in the global human discourse. Planners, engineers, architects, developers, and politicians are part of that discourse and work, to some degree, to promote communities that are indeed more sustainable. But none of these specialists or professionals has resilience and sustainability as their *first* priority or their primary area of expertise. They may make plans, design buildings, develop communities, and enact policies that may reduce vulnerability, enhance resilience, and contribute to sustainability, but their respective "professions" are to plan, design, build, develop, or legislate and hazard risks (natural or technological) are secondary concerns in their broader work. These secondary concerns are often mandated by legislation or policy, but even where they are voluntarily present for all of the right reasons, they are balanced against other and sometimes contradictory concerns. Indeed, many of the competing influences on their work may contribute to outcomes that, wittingly or unwittingly, contribute to the creation or toleration of hazard risks that are best avoided. Taking responsibility for disasters, and designing resilient and sustainable communities, requires another profession in the mix of specialists who influence the direction of community development, one that specializes in the identification and reduction of hazard risks as a first priority. It requires emergency management, especially hazard mitigation.

In its most generic sense, we have said that hazard mitigation reduces exposure to hazard risks. But it must, to the extent that hazard resilience is linked to sustainability, do more than reduce risk for those living in hazardous environments. Mitigation is more than flood embankments or levee systems. To achieve sustainability from an emergency management perspective also means that communities must choose where, when, and how development proceeds. It means keeping new development away from vulnerable locations. Most of the existing vulnerable locations are, perhaps, the result of historical decisions that cannot be reversed or addressed until redevelopment takes place. But even in less vulnerable locations, human decisions and actions may create new risks that threaten sustainability. Hazard mitigation must thus, as Gilbert White's insight suggests, be broadly designed to understand, respect, and protect the natural order by intelligently accommodating ourselves to its inevitable extremes. Emergency management must emphasize the nonstructural adaptations that accommodate nature and reduce risks with as much vigor and expertise as it does the structural techniques that may make risky environments more resilient. A reliance on structural mitigation techniques alone does not produce resilient communities (i.e., remember levees may only postpone a disaster) or promote long-term sustainability.

The emphasis on hazard mitigation as being essential for resilient and sustainable communities is not by any means to suggest that the response or recovery phases of emergency management are any less connected to resilience and sustainability. Resilient and sustainable communities should be able to *respond more efficiently and recover more quickly* when disasters do occur. It must be remembered that hazard mitigation cannot prevent all disasters. Even communities that have prioritized hazard mitigation are not immune to natural disasters. Dramatic geophysical events and/or limited resources available for or committed to hazard mitigation often combine to make the elimination or reduction of disaster impacts and costs impossible. Thus, the development of response resources and capacities is essential, but these must be connected to concerns about community resilience and sustainability as well. If critical infrastructure in a disaster area is not resilient, if response agencies are incapacitated by this, if emergency facilities, shelters, services, and government infrastructure are inaccessible during a disaster, then first response is unable to contribute to resilience. This puts additional pressure on the community to organize its own response to the disaster. In some cases, the immediate response may have to come from the victims, in which case a strategy to make them assets rather than mere victims may be a recommended part of predisaster planning and preparedness efforts. Thus, disaster response must be resilient and contribute to the resilience of the community when it is faced with disaster occurrence and its anticipated impacts.

Given that disasters happen even under the best of circumstances, the intelligent implementation of forward-thinking recovery strategies in the aftermath of an occurrence is a necessary part of building resilient and sustainable communities. Recovery is more than simple cleanup and restoration operations to get a community back on its feet. Often in the aftermath of a disaster, it is said by community leaders that the goal is to return the community to normal. But if the damage has been significant, and if it was in part caused by development decisions that created vulnerabilities, returning to normal is the last thing that needs to be done. The previous normal was wanting to the extent that vulnerabilities were either excessive or avoidable. Recovery is often the perfect disaster phase to assess and reduce vulnerabilities and to build more hazard resilience into the community. If restoration is the primary objective, this may not address the root causes of avoidable disaster damages and will only condemn the community to a repeated cycle of disaster-damage-repair-disaster-damage-repair. The designing of resilient and sustainable communities requires the inclusion of hazard mitigation and vulnerability reduction into the recovery phase of disasters.

Hazard resilience and sustainability are, like disaster costs, "largely acts of man." They are, like many of the disasters our communities experience, influenced or created by human design. Resilience and sustainability of course must be the work of entire communities and all of their professions. But hazard resilience in particular, as a contributing element to sustainability, is work that requires as a critical part of the nexus a fully developed emergency management profession. Planners, developers, architects, engineers, and others will confront hazard resilience and hazard mitigation only when required to do so by legislation usually. Emergency managers, in their roles as they relate to hazard mitigation planning, response preparedness, disaster response, and disaster recovery, are the ones primarily responsible for hazard resilience. It is their professional mission, we might say, as they work with communities to build mitigation capacity that contributes to the goal of sustainability.

Hazard resilience, as a necessary component in the building and maintaining of sustainable communities, is the primary mission of emergency management across all disaster phases, and it is the goal of each specialty within the emergency management profession. Remember also that hazard-resilient communities were said, in Chapter 2, to be characterized not only by technical hazard mitigation efforts but also by efforts to promote sustainable economic development, protect and enhance the natural resources, and ensure a better quality of life for its citizens. This ties the emergency manager and the emergency management profession to the network of sustainability in its broadest sense. The necessity of hazard resilience as part of building and maintaining sustainable communities makes emergency management a sustainability profession. It also makes emergency management a necessary profession.

EMERGENCY MANAGEMENT:
A SUSTAINABILITY PROFESSION

In the discussion of ethics and emergency management in Chapter 3, it was said that, to the extent that emergency management may be viewed as a sustainability enterprise, its core values and ethical dimensions would be clarified. Human history is in no small part defined by the never-ending effort to confront and adapt to the devastation caused by catastrophic events such as war, pestilence, and disaster. The anticipation of and management and/or reduction of risks that may contribute to such devastation is a logical necessity for sustainable communities. This is a necessity of which is born emergency management as a required societal function and profession.

Emergency managers know, to the extent that they have mastered the lessons of hazard mitigation or assessed the varied impacts of disasters, that disaster damages are not things that happen to a community while it is busy making other plans. They are instead most frequently the product of the plans it is busy making. The disaster does not define or create the situation of a community in relationship to hazard resilience and sustainability, but rather the decisions a community makes before the disaster hits create or harm its resilience and to a large extent predetermine its situation. The good news is, from an emergency management perspective, people and communities can adapt, learn from disaster experiences, and design hazard resilience, and excessive disaster losses or costs are not an inevitability as a result of disaster exposure. Hazard threats and potential losses associated with a disaster occurrence coexist with the human capacity to confront circumstances, risks, and vulnerabilities in ways that are proactive and adaptive. But this human capacity needs development and nurturing. It needs a clearly defined and vital emergency management profession.

Emergency managers know that the human capacity to develop hazard-resilient communities that promote sustainability is real. Sustainable hazard mitigation is, as they have learned over the past decades, *a choice that can be made* and a choice that can make a positive difference. But emergency managers also know that the real possibility of making decisions that lead to sustainable hazard resilience is no guarantee that they will be made. In too many cases, the decisions made by communities are the product of influences that work against sustainable hazard mitigation. This may be the result of choices or decisions driven by economic interests or financial concerns. Choices may be limited by public budgetary constraints or lack of competencies required to promote sustainable hazard mitigation as a priority in the competition with more immediate concerns on a community's agenda. Also, and perhaps most importantly, decisions or choices may be adversely shaped by a community's lack of commitment to develop the human capacity to confront circumstances, risks, and vulnerabilities in ways that are proactive and adaptive. Emergency managers know, or should know more precisely than others in any community, that the absence of such a commitment shared by all public and private entities in the community contributes to future community vulnerabilities and designs new and more costly disaster impacts.

Emergency managers know that more important than the decision to implement or not implement a specific hazard mitigation strategy is the general knowledge of mitigation or resilience resources that can (and should) be evaluated together with the assessments of risks and vulnerabilities to inform the identification of community development goals that are compatible with hazard resilience and sustainability. When a community informs its development goals thusly, it creates a culture of sustainability. A culture of sustainability requires, as a necessary component, hazard resilience. A sustainability culture logically enhances a community's ability to adapt to, deal with, and manage exposure to hazards and reduce the impact of any disasters. This implies taking responsibility at a societal level for the planning, policies, and actions necessary to prioritize hazard resilience. Taking responsibility increases substantially the confidence a community can have that the integration of engineering, lifeline, disaster planning, economic development, and all other policy initiatives contribute to resilience and sustainability. Now, and in light of all of this, emergency managers also know that they can't be the only ones taking responsibility. "It takes a village" (i.e., taking responsibility is a community-wide endeavor). But emergency managers also know, as should the entire community, that taking responsibility for disasters in any community requires what emergency managers know and what they do as an integral and absolutely essential component. Emergency managers and what they know are necessary if a community wants to take responsibility. It is the necessity of emergency managers and their role in a community's taking responsibility for disasters that should ultimately define the purpose of an emergency management profession.

As we noted in Chapter 3, emergency management is perhaps best defined by its focus on sustainability and its adherence to it as the social value to be served by the application of its knowledge and the meeting of the forward-thinking responsibilities imposed by it. This is to suggest that hazard resilience, as a necessity for sustainable communities, is the defining principle of the profession. It can be said of course that hazard resilience is the work of an entire community, but the work must be carefully planned and implanted. For that to happen, someone (or some entity) must have responsibility for leading the way. This is where emergency managers (and their putative profession ideally) fit into the mix. They and their profession have unique knowledge and experience that bring them front and center in the community-wide commitment to build hazard resilience into the social fabric as a necessary part of creating and maintaining sustainable communities. Emergency management, to such degree that it evolves into a more fully defined and advanced profession, is thus a *necessary* profession. What it is necessary for makes emergency management first and foremost a sustainability profession. This has certain implications that ideally should influence the future evolution of the profession.

We noted at the beginning of this book that the transformation of emergency management from trade to profession, from occupation to vocation, required that more attention be paid to the mental or intellectual foundations than to the manual work or specific technical functions that come with the job. As we have discussed the evolution of the work emergency managers do, its all-hazards orientation, its expansion from a first-response orientation to include hazard mitigation as a critical component, its managerial complexities, its evolving principles and ethical dimensions, the challenges and ever-changing vulnerability profiles presented by natural and human-created hazards, and the connectivity between it and the broader planning, policy, and development activities of the communities it serves, the mental or intellectual foundations of a profession have in fact been laid out for us.

Increasingly it has become a fait accompli that emergency management is, in all of its phases and specialties, a critical part of developing a community's adaptive capacity in the face of multiple hazard threats. Adaptive capacities change over time as do the array of hazards communities must manage. Priorities, national and local, change. Terrorism overshadowed natural disasters as a national focus in the immediate aftermath of 9-11. Hazard landscapes and disaster footprints change. Environmental degradation and global climate change have altered the course of recurring threats, and they have created new ones. Communities change. Demographically, economically, and physically, community needs and expectations change as do the impacts (positive and negative) of population growth, worsening or improving economic conditions, and the spread of the constructed environment. What *never changes* in communities that are sustainable is the ongoing need for the assessment of risks and vulnerabilities and the need to incorporate them into community planning, the need for the implementation of risk reduction or mitigation strategies, the need to evaluate and improve response capacities, and the periodic need to rebuild or recover and improve resilience after a disaster occurrence. The specifics of these functions related to planning, mitigation, response, and recovery may change, but the need for these functions remains constant. For the practicing emergency manager, keeping apace of the changes in priorities, hazard landscapes, and communities, and the escalating number of hazard risks and vulnerabilities that accompany many of these changes, is daunting and, perhaps on some level, depressing.

Emergency managers may feel that the pace of changing risk and vulnerability profiles is too rapid for them to stay ahead of as they focus on the technical aspects of their work and feel perhaps a bit overwhelmed. In the face of a changing climate, the creation of new risks associated with human technological and industrial endeavors, the reluctance of corporations, businesses, policymakers, or communities to take responsibility for disasters, emergency managers may ask themselves, "Who are we to think we can manage or be responsible for all of this?" The feeling that things are moving faster than they can be managed is understandable. But if one takes a deep breath and reflects on the road emergency management has traveled over the past several decades, there is also reason for hope that better times are possible. Generally speaking, it is indeed true that emergency management is, in all of its phases, increasingly perceived as a critical and necessary part of developing a community's adaptive capacity in the face of multiple hazard threats. Generally speaking, especially with respect to hazard mitigation and the importance of hazard resilience, the potential for effective action has been expanded with the experience, tools, and knowledge acquired over the past few decades.

Emergency managers have been increasingly trained and are more and more experienced in the technical functions associated with planning, disaster response, recovery, and pre- and postdisaster mitigation. But the training and experience are largely skills oriented and often a reactionary mix of things designed to respond to specific laws or policies aimed at specific hazards and responses. Their expertise, such as it is, is not holistic so much as it is an accumulation of skills and knowledge related to hazards and disasters. Even programs in higher education settings (which theoretically provide some of the necessary theoretical development and analytical skills) tend not to be holistic. Some might think that formal educational settings provide more theory and less training, but the reality is (as noted in Chapter 1), programs in higher education also tend to be a reactionary accumulation of topics related to hazards and disasters. What is most needed for the promotion of an emergency management profession is a linkage of the expertise and skills of its practitioners to a vision of the broader purpose (i.e., value) to be served by the work. The end product of emergency management, as a profession, must be more than the sum of its practitioners' work.

The thrust of this chapter and of this book has been that the end product of emergency management must be understood as fundamentally connected to all facets of community life in a coordinated effort with all relevant actors, public and private, to promote sustainability. This means that in addition to the technical skills that relate to each disaster phase in the emergency management cycle, emergency managers must bring knowledge and a perspective to the table that are relevant to the broader task of sustainable community development. This knowledge or perspective must be what integrates all of the technical components and skills of the profession and directs them toward the fulfillment of a broader purpose (i.e., value). Sustainability is that broader purpose or value. Hazard resilience, as a necessity for sustainable communities, is the defining principle of the profession. Developing a community's adaptive capacity in the face of multiple hazard threats is the emergency management profession's primary mission. This is what it means when we say emergency management is a sustainability profession.

The work, the training, the development of programs in higher education, the development of credentialing instruments, and the vertical organization of the profession and its ranks must (as they continue to evolve) all be integrated into the sustainability framework. Emergency managers are increasingly faced with problems (natural and human made) they have never before confronted, and they are expected to comprehend complex physical and social systems. They are increasingly expected to offer long-term solutions for recurring problems. Obviously, technical, analytical, and critical thinking skills have never been more important. Beyond these skills, however, it is critically important to work within a worldview that comprehends the connections among hazard threats, disasters, and sustainability. This theme (the linkage of emergency management to sustainability) must orient all of the professional skill development, educational curriculum, and work of the profession generally. If we take Gilbert White's insight (disaster damages are "acts of men") as a starting point, taking responsibility for disasters is a necessary prerequisite for sustainable communities. This is what makes emergency management, as a profession, necessary and perhaps even intelligible.

What do emergency managers do? The book began with this question. We have seen that emergency managers anticipate disasters and take preventive and preparatory measures to build disaster-resistant and disaster-resilient communities; use sound risk management principles (hazard identification, risk analysis, and impact analysis) in defining priorities and assigning resources; sustain broad relationships among individuals and groups to create trust, build consensus, and promote resilience; coordinate the activities of public and private stakeholders to achieve a common purpose; use creative approaches in solving disaster problems; and value a scientific and knowledge-based approach with respect to education, training, ethical practice, and public stewardship. These are, of course, the recognized principles of emergency management we discussed in Chapter 1. These are the things emergency managers do. All of these things say one basic thing. Emergency managers, on behalf of an entire community committed to resilience and sustainability, take responsibility for disasters.

As a sustainability profession, emergency management is an embodiment of a community's commitment to develop its adaptive capacity in the face of multiple (natural or human-made) hazard threats. The profession is premised on the need to respect and protect the natural order by intelligently accommodating ourselves to its inevitable extremes and the need to identify and manage the hazard risks produced by our own actions. Communities that do not meet these two needs cannot be sustainable. Emergency management, as a profession, is a prerequisite for sustainable communities. The profession knows that disaster costs and hazard resilience are both "largely acts of men," and sustainability requires us to take responsibility for our acts and choose resilience over disasters by design. The profession knows one thing above all and acts proactively on that knowledge. In Gilbert White fashion, emergency management begins with one basic but critical insight. Disaster damages are not, as we have said throughout this book, things that happen to a community while it is busy making other plans. They are the product of the plans it is busy making. Emergency management is first and foremost the taking of responsibility for disasters and the planning for sustainability. This is the work of an entire community, but it requires a "profession" to guide its efforts.

RECOMMENDED READING

ALGA. (2006). Local government land use planning and risk mitigation: National research paper. Available online: http://www.alga.asn.au/policy/emergman/pdf/LGLUP.pdf.

Baker, E., & McPhee, J. (1975). *Land use regulations in hazardous areas: A research assessment.* Boulder: Natural Hazards Research and Applications Information Centre, University of Colorado.

Beatley, T. (1998). The vision of sustainable communities. In R. J. Burby (Ed.), *Cooperating with Nature: Confronting Natural Hazards with Land-Use Planning for Sustainable Communities* (pp. 233–262). Washington, DC: Joseph Henry Press.

Becker, J. S., & Saunders, W. S. A. (2007, March). Enhancing sustainability through pre-event recovery planning. *Planning Quarterly,* 14–18.

Bell, R., Hume, T., & Todd, D. (2002, July). Planning on rising sea level? *Planning Quarterly,* 13–15.

Berke, P. R. (1995). Natural hazard reduction and sustainable development: A global assessment. *Journal of Planning Literature, 9*(4), 370–382.

Berke, P. R. (1998). Reducing natural hazard risks through state growth management. *Journal of the American Planning Association, 64*(1), 76–87.

Brody, S. D. (2003). Are we learning to make better plans? A longitudinal analysis of plan quality associated with natural hazards. *Journal of Planning Education and Research, 23,* 191–201.

Brody, S. D., Zahran, S., Maghalel, P., Grover, H., & Highfield, W. E. (2007). The rising cost of floods: Examining the impact of planning and development decisions on property damage in Florida. *Journal of the American Planning Association, 73*(3), 330–345.

Bryant, E. A., Head, L., & Morrison, R. J. (2005, February 2–5). Planning for Natural Hazards – How Can We Mitigate the Impacts? In R. J. Morrison, S. Quin, and E. A. Bryant (Eds.), *Planning for Natural Hazards – How Can We Mitigate the Impacts?* Proceedings of a Symposium, University of Wollongong, GeoQuEST Research Centre, 1–11. Available at: http://ro.uow.edu.au/scipapers/59.

Burby, R. J., with Beatley, T., Berke, P. R., Deyle, R. E., French, S. P., Godschalk, D. R., Kaiser, E. J., Kartez, J. D., May, P. J., Olshansky, R., Paterson, R. G., & Platt, R. H. (1999, Summer). Unleashing the power of planning to create disaster resistant communities. *Journal of the American Planning Association,* 247–258.

Burby, R. J., Deyle, E. W., Godschalk, D. R., & Olshansky, R. B. (2000). Creating hazard-resilient communities through land-use planning. *Natural Hazards Review, 1,* 99–106.

Button, G. (2010). *Knowledge and Uncertainty in the Wake of Human and Environmental Catastrophe.* Walnut Creek, CA: Left Coast Press.

Campanella, T. J. (2006). Urban resilience and the recovery of New Orleans. *Journal of the American Planning Association, 72*(2), 141–146.

Deyle, R. E., Chapin, T. S., & Baker, E. J. (2008). The proof of the planning is in the platting: An evaluation of Florida's hurricane exposure mitigation planning mandate. *Journal of the American Planning Association, 74*(3), 349–370.

Ellis, S., Kanowski, P., & Whelan, R. (2004). *National Inquiry on Bushfire Mitigation and Management*. Canberra: Commonwealth of Australia.

Geis, D., & Kutzmark, T. (1995, August). Developing sustainable communities: The future is now. *Public Management*, 4–13.

Glavovic, B. C. (2008). Sustainable coastal communities in the age of coastal storms: Reconceptualising coastal planning as "new" naval architecture. *Journal of Coastal Conservation, 12*, 125–134.

Glavovic, B. C. (2013). Realising the promise of natural hazards planning: An Australasian perspective. *Australasian Journal of Disaster and Trauma Studies*.

Godschalk, D. R. (2003). Urban hazard mitigation: Creating resilient cities. *Natural Hazards Review, 4*(3), 136–143.

Godschalk, D. R., Beatley, T., Berke, P., Brower, D. J., & Kaiser, E. J. (1998). *Natural Hazard Mitigation: Recasting Disaster Policy and Planning*. Washington, DC: Island Press.

Haddow, G., Bullock, J. A., & Haddow, K. (2009). *Global Warming, Natural Hazards, and Emergency Management*. Boca Raton, FL: CRC Press.

Haque, C. E., & Etkin, D. (2007). People and community as constituent parts of hazards: The significance of societal dimensions in hazards analysis. *Natural Hazards, 41*, 271–282.

Irazábal, C., & Neville, J. (2007). Neighbourhoods in the lead: Grassroots planning for social transformation in post-Katrina New Orleans? *Planning Practice and Research, 22*(2), 131–153.

Jacobson, M. (2005, March). NZCPS coastal hazard policies. *Planning Quarterly*, 6–8.

King, D. (2006). Planning for hazard resilient communities. In D. Paton and D. Johnston (Eds.), *Disaster Resilience: An Integrated Approach* (pp. 288–304). Springfield, IL: Charles C Thomas.

Knowles, S. G. (2011). *The Disaster Experts: Mastering Risk in Modern America*. Philadelphia: University of Pennsylvania Press.

Mileti, D. S. (Ed.). (1999). *Disasters by Design: A Reassessment of Natural Hazards in the United States*. Washington, DC: Joseph Henry Press.

Mileti, D. S., & Gailus, J. L. (2005). Sustainable development and hazards mitigation in the United States: Disasters by Design revisited. *Mitigation and Adaptation Strategies for Global Change, 10*, 491–504.

Olshansky, R. B. (2001). Land use planning for seismic safety: The Los Angeles County Experience, 1971–1994. *Journal of the American Planning Association, 67*(2), 173–185.

Olshansky, R. B. (2006). Planning after Hurricane Katrina. *Journal of the American Planning Association, 72*(2), 147–153.

Olshansky, R. B., Johnson, L. A., Horne, J., & Nee, B. (2008). Planning for rebuilding New Orleans. *Journal of the American Planning Association, 74*(3), 273–287.

Olshansky, R. B., & Kartez, J. (1998). Managing land-use to build resilience. In R. J. Burby (Ed.), *Cooperating with Nature: Confronting Natural Hazards with Land-Use Planning for Sustainable Communities* (pp. 167–201). Washington, DC: Joseph Henry Press.

Paton, D., & Johnston, D. (Eds.). (2006). *Disaster Resilience: An Integrated Approach*. Springfield, IL: Charles C Thomas.

Puszkin-Chevlin, A., Hernandez, D., & Murley, J. (2006/2007). Land use planning and its potential to reduce hazard vulnerability: Current practices and future possibilities. *Marine Technology Society Journal, 40*(4), 7–15.

Quarantelli, E. L. (1998). *What Is a Disaster? Perspectives on the question.* London: Routledge.

Rodriquez, J., Quarantelli, E., & Dynes, R. (Eds.). (2006). *Handbook of Disaster Research.* New York: Springer.

Schwab, A. J., Eschelbach, K., & Brower, D. J. (2007). *Hazard Mitigation and Preparedness: Building Resilient Communities.* Hoboken, NJ: Wiley.

Smith, G. (2008). Planning for sustainable and disaster resilient communities. In J. Pine (Ed.), *Hazard Analysis.* Washington, DC: Taylor & Francis.

Smith, G., & Wenger, D. (2006). Sustainable disaster recovery: Operationalizing an existing agenda. In J. Rodriquez, E. Quarantelli, and R. Dynes (Eds.), *Handbook of disaster research.* New York: Springer.

Van Aalst, M. K. (2006). The impacts of climate change on the risk of natural disasters. *Disasters, 30*(1), 5–18.

Van Roon, M. (2003, December). Managed flooding? *Planning Quarterly,* 9–12.

Varely, A. (Ed.). (1994). *Disasters, Development and Environment.* London: John Wiley.

WCDR. (2005, January 18–22). *Building the Resilience of Nations and Communities to Disasters.* World Conference on Disaster Reduction, Proceedings of the Conference, Kobe, Hyyogo, Japan. Geneva: United Nations.

White, G. F. (1936). Notes on flood protection and land use planning. *Planners Journal, 3*(3), 57–61.

White, G. F. (1945). *Human Adjustment to Floods* (Research Paper No. 29). Chicago, IL: University of Chicago Press.

White, G. F., Calef, W. C., Hudson, J. W., Mayer, H. M., Sheaffer, J. R., & Volk, D. J. (1958). *Changes in Urban Occupance of Flood Plains in the United States* (Research Paper No. 77). Chicago, IL: University of Chicago Press.

Wisner, B., Blaikie, P., Cannon, T., & Davis, I. (2004). *At Risk. Natural Hazards, People's Vulnerability and Disasters.* London: Routledge.

Wright, K., Becker, J., & Saunders, W. (2009). Pre-event Recovery Planning for Natural Hazards. *Tephra, 22,* 49–54.

EPILOGUE

Any book discussing the various aspects of emergency management, and especially one that addresses the definition of an emergency management "profession," is an incomplete work. In part this is because the defining of the profession is a work in progress. In part it is because of the dynamic nature of the subject matter. The ongoing and regular cycle of disaster occurrences and the constantly changing risk and vulnerability profiles of human populations around the world make for a rapidly changing or evolving landscape so to speak. Beyond the larger conceptual framework guiding such a discussion is the ever-present reality that disasters happen and disaster damages occur. This reality makes the larger conceptual conversation more important to the extent that it is aimed at taking responsibility for disasters as a necessary prerequisite for sustainable communities. It also invites an epilogue, a final remark that comments on or summarizes the "main action" or "plot" of an ongoing conversation. Without any pretensions of eloquence or profundity, a few concluding remarks do seem to be quite in order.

As these remarks are being composed, two events over a four-month period at the end of 2012 seem to be reminders of the theme that disasters are in part a matter of human design. They may also be harbingers of the future. The first of these is Hurricane Isaac (August 21 to September 1, 2012). A modest category 1 hurricane, Isaac directly impacted Puerto Rico, Hispaniola, Cuba, the Bahamas, Florida, Louisiana, Mississippi, and Alabama. An estimated $3 billion in damages and forty-four fatalities lay in its wake. Accompanying this natural disaster were some human-made hazards. Refinery accidents in the Gulf during Hurricane Isaac included overturned chemical tanks, oil sheens, runoff in Plaquemines Parish wetlands, thousands of pounds of chemicals dumped by Exxon's Chalmette Refining, and many other smaller chemical accidents. According to a number of media reports, many of these accidents and leaks were preventable. This typical narrative reminds us that natural and human-made hazards often combine to influence disaster impacts. The beat goes on we might say. More dramatic with respect to future implications may be the event known as Superstorm Sandy.

Sandy began in the Caribbean on October 19, 2012. It began as a tropical depression and within six hours became a tropical storm. It was upgraded to hurricane status on October 24th. As it tore through the Caribbean, Sandy caused more than $300 million in damages. Sandy swept across the Bahamas, briefly weakened to a tropical storm on October 27th, and then strengthened again to become a Category 1 hurricane as it turned north toward the U.S. coast. Hurricane Sandy made landfall in the United States on the evening of October 29th near Atlantic City, New Jersey. As it wreaked havoc on the U.S. Northeast, Sandy resulted in 253 deaths and caused estimated damages of $65 billion. The impact of the storm, its landing in the American Northeast (i.e., the media center of the nation), its size and complexity (i.e., tropical storm winds extending 820 miles, its pure kinetic energy for storm surge and wave destruction potential reaching 5.8 on a scale of 0 to 6, heavy snows in West Virginia, etc.), and its costs with respect to damages and losses combined to almost immediately change the national conversation somewhat with respect to natural disasters and climate change.

In the immediate aftermath of Sandy, the anticipation of future "Sandys" seemed to resonate. Questions and confusion from some people whose homes and businesses were destroyed by Hurricane Sandy included trying to figure out whether they should rebuild or even could rebuild and, if so, how. Much of the official discussion in the aftermath of Sandy includes a sense of heightened seriousness about the need to rebuild devastated areas with smarter designs. The talk is not just about rebuilding, but rebuilding smarter in anticipation of future disasters of a serious magnitude. This is a classic case of mitigation discussion and decision making during the recovery stage to be sure but with an added sense of urgency in the context of a changing climate.

Hurricane Isaac and Superstorm Sandy are, of course, two in an array of natural disasters that occur on a recurring cycle. A mere listing of natural disasters in the United States throughout the twelve months of 2012 attests to the unending challenges posed by natural extremes.

The first eight months of 2012 in the United States were dotted with such occurrences (e.g., a Pacific Northwest Snowstorm in January, a rare out-of-season tornado outbreak in January in the American Southeast, an April Fool's Day blizzard in New England, New York, and New Jersey, a series of huge wildfires in the American West through the summer, and a number of severe storms and floods). The summer of 2012 saw record warm temperatures across the country, and a majority of U.S. counties, more than 60% of them, experienced severe drought conditions. The tendency is to consider all of these events as "normal." But Superstorm Sandy has suggested, to many at least according to what one can read in recent press reports, that "normal" isn't what it used to be. State governments are seeking funding and strategies to head off future disasters as climate scientists continue to predict the potential for more bad storms that exceed our previous expectations.

Natural disasters are recurring events familiar to most regions of the United States. They do indeed occur in the normal course of events. But in both 2011 and 2012, it seemed that nature was pummeling the United States with more weather extremes. Unprecedented heat, devastating drought contributing to significant crop losses, more deadly tornadoes leveling towns, massive rivers overflowing, and rising disaster-related costs became more frequent visitors to the evening news and cable news networks. Again, extreme weather and climate events – such as drought, heavy rain, and heat waves – are indeed a natural part of the earth's climate system. But as the climate changes, many of these extremes may occur outside the historical range, resulting in increased societal and environmental vulnerabilities. According to the U.S. Climate Change Science Program, most of North America has been experiencing more unusually hot days and nights, fewer unusually cold days and nights, and fewer frost days. Storms and extreme weather events have become more frequent and intense, and droughts are becoming more severe. Weather and climate extremes are expected to become more frequent during the 21st century. Recent reports seem only to confirm this conclusion.

In January 2013, the National Climate Data Center in Asheville, NC, reported that 2012 had been the warmest year on record in the United States. Particularly significant was the ratio of heat records to cold records. 2012 saw 362 all-time high temperatures across the nation and zero all-time record lows. In addition, 61% of the nation experienced moderate to exceptional drought conditions (2012 was the driest year in the United States since 1988). 2012 also included eleven natural disasters in the United States that reached the $1 billion threshold in damages and losses, culminating in the dramatic impact of Superstorm Sandy in October/November.

With the sheer number of recurring and expected natural disasters and their costs, with the sheer number of human-made hazards that together with natural hazards require anticipatory and proactive policies and procedures, with the constantly and more rapidly changing risk and vulnerability profiles associated with a changing climate, and with the implications all of this may have for resilience and sustainability, it is difficult to imagine a more universally needed and valuable public service than that associated with emergency management. It is also difficult to find a necessary and valuable public service that is less discussed and perhaps less understood by the communities it serves. Some of this, as we have seen, is a function of the fact that most people do not pay attention to hazards and disasters until an occurrence brings it to their immediate attention, often in the midst of a tragedy that was avoidable. But, as emphasized throughout our discussion in this book, much of it may be attributed to the imprecise or incomplete definition and development of emergency management as a profession.

The evolving technical excellence of emergency management in all of its phases is beyond dispute. The mainstreaming of emergency management in relationship to the critical concerns associated with resilience and sustainability, although somewhat tentative in practice and uneven, is certainly something that is in motion. But emergency management's professional identity is not yet fully formed, and its voice, as a profession, is relatively inaudible and unpersuasive on broader issues that relate to or impact its work and, quite frankly, that require its perspective as it relates to the building and maintaining of sustainable communities. One of the main contentions of this book has been that the mainstreaming of emergency management as a sustainability profession into the broader and inclusive task of sustainable community development is both necessary and more urgently required. But it is worth asking whether the development of such a professional identity and the mainstreaming of emergency management into community development decision making can really make a difference in the end. Even the most fully evolved professions are easily ignored by policymakers and citizens. But then, not all professions are positioned to fill a critical void in the public dialogue.

The issue of climate change may be the perfect example of the void to be filled by an emergency management profession. The climate, as we have suggested, is changing before our eyes. There have been more record high temperatures, and 2012 brought the worst drought in decades. More Arctic sea ice is melting and more rapidly than at any time in recorded history. More destructive wildfires, more extreme weather-related events, and the ongoing scientific research all suggest that the impact of climate change with respect to sustainability has perhaps never required more urgent discussion. Indeed, the experiences of this past year culminating in Superstorm Sandy suggest that we are possibly on the forefront of a climate disaster. Yet in a national election year (2012) in the United States, the issue of climate change was not seriously discussed. One of the national political parties ridiculed the science, and the other seemed, aside from an occasional mention of it, to be content to avoid the topic. Neither appeared serious about resolving what may be the most serious issue of our time. All the science in the world did not, it appears, move this discussion forward much in the American political discourse. So again, why should we assume that a vigorous mainstreaming of emergency management as a sustainability profession would make a difference on this or any other issue related to the building of resilient and sustainable communities? The premise of this book is, in fact, it might make more of a difference than we think.

The mainstreaming of emergency management into a discussion about climate change, the examination of that issue from an emergency management perspective with a focus on assessing changing risks and vulnerabilities to our communities, its impacts on our economies and on public health, and the costs of the potential disaster damages that may be related to it may attract far more urgent attention than all the scientists and politicians in the world have generated to date. Indeed, recent events (e.g., Sandy) seem to be altering the conversation as we speak. The tilting of the topic and the examination of it through a new prism seems possible, and it can, perhaps, make all the difference. It may be this missing emergency management perspective that is the missing link so to speak. Likewise, involvement in such a conversation and assessment can only better prepare practicing emergency managers to deal more effectively with new challenges already being imposed on them and their work by a changing climate. That looks pretty much like a win-win situation with respect to hazard resilience and the broader goal of sustainability.

The journey from trade to profession is still ongoing for emergency management, and there is no doubt that it has many miles to go. But the completion of that journey, to the extent that it defines its objectives and knows its destination to be connected to sustainability, will make a great difference indeed.

INDEX